# WiMAX Systems: Current Technologies

# WiMAX Systems: Current Technologies

Brett Rice

**W**ILLFORD **P**RESS
www.willfordpress.com

Published by Willford Press,
118-35 Queens Blvd., Suite 400,
Forest Hills, NY 11375, USA

ISBN: 978-1-64728-336-0

**Cataloging-in-Publication Data**

WiMAX systems : current technologies / Brett Rice.
    p. cm.
Includes bibliographical references and index.
ISBN 978-1-64728-336-0
1. IEEE 802.16 (Standard). 2. Wireless metropolitan area networks--Standards.
3. Ethernet (Local area network system). I. Rice, Brett.
TK5105.55 .W56 2022
004.62--dc23

For information on all Willford Press publications
visit our website at www.willfordpress.com

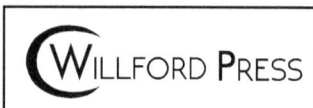

WILLFORD PRESS

# TABLE OF CONTENTS

It is with great pleasure that I present this book. It has been carefully written after numerous discussions with my peers and other practitioners of the field. I would like to take this opportunity to thank my family and friends who have been extremely supporting at every step in my life.

WiMAX stands for Worldwide Interoperability for Microwave Access. It is a set of wireless broadband communication standards which are based on IEEE 802.16. It provides media access control and multiple physical layer options. WiMAX 2+ is the latest version and it is a backwards-compatible transition from previous generations. Subscriber stations are the devices that provide connectivity to WiMAX network. It is a long-range system as it covers many kilometers. It uses a QoS mechanism that is based on connections between user device and the base station. WiMAX provides portable mobile broadband connectivity through various devices. It acts a wireless alternative to digital subscriber line and cable. It makes triple-play service offerings possible. Network of WiMAX can be accessed through USB when connected to dongle. WiMAX is an upcoming field of science that has undergone rapid development over the past few decades. This book is compiled in such a manner, that it will provide an in-depth knowledge about the theory and practice of WiMAX. Those in search of information to further their knowledge will be greatly assisted by this textbook.

The chapters below are organized to facilitate a comprehensive understanding of the subject:

Chapter – Introduction

WiMAX stands for Worldwide Interoperability for Microwave Access. It refers to a family of wireless broadband communication standards based on the IEEE 802.16 set of standards, which provide multiple physical layers and media access control options. This is an introductory chapter which will briefly introduce about WiMAX.

Chapter – Wireless Network and its Types

Wireless network is a type of network that uses radio signals for communication among computers and other network devices. Wide area network, local area network and per sonal area network are its three major types. This chapter has been carefully written to provide an easy understanding of the wireless network and its types.

Chapter – Mobile WiMAX

Mobile WiMAX is a technical wireless system that enables convergence of mobile and fixed broadband networks through a common wide area network. Mobile IP, 3G IEEE family, 4G IEEE family, access service network gateway, etc. are some of the concepts that fall under mobile WiMAX. All these concepts of mobile WiMAX have been carefully analyzed in this chapter.

Chapter – Protocol Layers and Topologies

WiMAX physical layers are of five types namely SC, SCa, HUMAN, OFDM and OFDMA. A point-to-multipoint topology is operated using LOS signal propagation for providing network access from one location to another. The topics elaborated in this chapter will help in gaining a better perspective about protocol layers and topologies.

Chapter – Wireless Display Technologies

Wireless display technologies enable the users to stream music, movies, photos, videos and applications without wires from a compatible computer to a compatible HDTV. Some of these technologies are Chromecast, Google cast, EZCast, Micracast, OpenFlint, etc. This chapter closely examines these wireless display technologies to provide an extensive understanding of the subject.

Chapter – Orthogonal Frequency Division Multiplexing

Orthogonal frequency division multiplexing is used for wideband digital communication through encoding digital data on multiple carrier frequencies. It has varied applications in digital television and audio broadcasting, wireless networks, power line networks, and 4G mobile communications. This chapter discusses orthogonal frequency division multiplexing in detail.

Chapter – Applications

WiMAX is used for a wide-range of applications such as broadband connections, cellular backhaul, hotspots, e-learning, wider metropolitan area network access to users, etc. This chapter closely examines the different applications of WiMAX for an extensive understanding of the subject.

**Brett Rice**

# Introduction

# CHAPTER 1

WiMAX stands for Worldwide Interoperability for Microwave Access. It refers to a family of wireless broadband communication standards based on the IEEE 802.16 set of standards, which provide multiple physical layers and media access control options. This is an introductory chapter which will briefly introduce about WiMAX.

## WIMAX

WiMAX is a standardized wireless version of Ethernet intended primarily as an alternative to wire technologies (such as Cable Modems, DSL and T1/E1 links) to provide broadband access to customer premises.

More strictly, WiMAX is an industry trade organization formed by leading communications, component, and equipment companies to promote and certify compatibility and interoperability of broadband wireless access equipment that conforms to the IEEE 802.16 and ETSI HIPERMAN standards.

WiMAX would operate similar to WiFi, but at higher speeds over greater distances and for a greater number of users. WIMAX has the ability to provide service even in areas that are difficult for wired infrastructure to reach and the ability to overcome the physical limitations of traditional wired infrastructure.

WIMAX was formed in April 2001, in anticipation of the publication of the original 10-66 GHz IEEE 802.16 specifications. WIMAX is to 802.16 as the WiFi Alliance is to 802.11.

WIMAX is:

- Acronym for Worldwide Interoperability for Microwave Access.

- Based on Wireless MAN technology.

- A wireless technology optimized for the delivery of IP centric services over a wide area.

- A scalable wireless platform for constructing alternative and complementary broadband networks.

- A certification that denotes interoperability of equipment built to the IEEE 802.16 or compatible standard. The IEEE 802.16 Working Group develops standards that address two types of usage models:
  - A fixed usage model (IEEE 802.16-2004).
  - A portable usage model (IEEE 802.16e).

## 802.16a

WiMAX is such an easy term that people tend to use it for the 802.16 standards and technology themselves, although strictly it applies only to systems that meet specific conformance criteria laid down by the WiMAX Forum.

The 802.16a standard for 2-11 GHz is a wireless metropolitan area network (MAN) technology that will provide broadband wireless connectivity to Fixed, Portable and Nomadic devices.

It can be used to connect 802.11 hot spots to the Internet, provide campus connectivity, and provide a wireless alternative to cable and DSL for last mile broadband access.

## WiMAX Speed and Range

WiMAX is expected to offer initially up to about 40 Mbps capacity per wireless channel for both fixed and portable applications, depending on the particular technical configuration chosen, enough to support hundreds of businesses with T-1 speed connectivity and thousands of residences with DSL speed connectivity. WiMAX can support voice and video as well as Internet data.

WIMAX developed to provide wireless broadband access to buildings, either in competition to existing wired networks or alone in currently unserved rural or thinly populated areas. It can also be used to connect WLAN hotspots to the Internet. WiMAX is also intended to provide broadband connectivity to mobile devices. It would not be as fast as in these fixed applications, but expectations are for about 15 Mbps capacity in a 3 km cell coverage area.

With WiMAX, users could really cut free from today's Internet access arrangements and be able to go online at broadband speeds, almost wherever they like from within a Metro-Zone.

WIMAX could potentially be deployed in a variety of spectrum bands: 2.3GHz, 2.5GHz, 3.5GHz, and 5.8GHz.

## Need of WiMAX

- WiMAX can satisfy a variety of access needs. Potential applications include extending broadband capabilities to bring them closer to subscribers, filling gaps in cable, DSL and T1 services, WiFi, and cellular backhaul, providing last-100 meter access from fibre to the curb and giving service providers another cost-effective option for supporting broadband services.

- WiMAX can support very high bandwidth solutions where large spectrum deployments (i.e. >10 MHz) are desired using existing infrastructure keeping costs down while delivering the bandwidth needed to support a full range of high-value multimedia services.

- WiMAX can help service providers meet many of the challenges they face due to increasing customer demands without discarding their existing infrastructure investments because it has the ability to seamlessly interoperate across various network types.

- WiMAX can provide wide area coverage and quality of service capabilities for applications ranging from real-time delay-sensitive voice-over-IP (VoIP) to real-time streaming video and non-real-time downloads, ensuring that subscribers obtain the performance they expect for all types of communications.

- WiMAX, which is an IP-based wireless broadband technology, can be integrated into both wide-area third-generation (3G) mobile and wireless and wireline networks allowing it to become part of a seamless anytime, anywhere broadband access solution.

Ultimately, WIMAX is intended to serve as the next step in the evolution of 3G mobile phones, via a potential combination of WiMAX and CDMA standards called 4G.

## WiMAX Forum

The WiMAX Forum is a wireless industry consortium with a growing number of members including many industry leaders. It has been set up to support and develop WiMAX technology worldwide, bring common standards across the globe to enable the technology to become an established worldwide technology.

One of the aims of the forum is to enable a standard to be adopted that will enable full interoperability between products. Learning from the problems of poor interoperability experienced with previous wireless standards, and the impact that this had on take up, the WiMAX Forum aims to prevent this from happening. Ultimately vendors will be able to have products certified under the auspices of the Forum, and then be able to advertise their products as "Forum Certified".

Although WiMAX technology will support traffic based on transport technologies ranging from Ethernet, Internet Protocol (IP), and Asynchronous Transfer Mode (ATM), the Forum will only certify the IP-related elements of the 802.16 products. The focus is on IP operations because this is the now the main protocol that is used.

## WiMAX Versions

Since its initial conception, new applications for WiMAX have been developed and as a result there are two "flavours" of WiMAX technology that are available.

The two flavours of WiMAX broadband technology are used for different applications and although they are based on the same standard, the implementation of each has been optimised to suit its particular application.

- 802.16d - DSL replacement: The 802.16d version is often referred to as 802.16-2004 and it is closer to what may be termed the original version of WiMAX defined under 802.16a. It is aimed at fixed applications and providing a wireless equivalent of DSL broadband data - often called WiMAX broadband. In fact the WiMAX Forum describes the technology as "a standards-based technology enabling the delivery of last mile wireless broadband access as an alternative to cable and DSL."

  802.16d is able to provide data rates of up to 75 Mbps and as a result it is ideal for fixed, DSL replacement applications as WiMAX broadband. It may also be used for backhaul where the final data may be distributed further to individual users. Cell radii are typically up to 75 km.

- 802.16e - Nomadic/Mobile: While 802.16/WiMAX was originally envisaged as being a fixed only technology, with the need for people on the move requiring high speed data at a cost less than that provided by cellular services and opportunity for a mobile version was seen and 802.16e was developed. This standard is also widely known as 802.16-2005. It currently provides the ability for users to connect to a WiMAX cell from a variety of locations, and there are future enhancements to provide cell handover.

  802.16e is able to provide data rates up to 15 Mbps and the cell radius distances are typically between 2 and 4 km.

## Competition

The competition with WiMAX, 802.16 depends upon the type or version being used. Although initially it was thought that there could be significant competition with Wi-Fi, there are other areas to which WiMAX is posing a threat.

- DSL cable lines: WiMAX is able to provide high speed data links to users and in this way it can pose a threat to DSL cable operators.

- Cell phone operators: As LTE was being developed and the initial roll-outs were taking place, cell phone operators saw the mobile version of WiMAX as a significant threat. It was even considered for adoption as the IMT 4G standard, but LTE was adopted as the standard, leaving WiMAX for fixed WiMAX braodband, last mile links and a variety of other point to point applications.

WiMAX technology has been deployed in many areas. Although initially seen as a candidate for 4G, its use is decreasing, although it is used as WiMAX broadband and also for last mile links.

## WiMAX Advantages

WiMAX is popular because of its low cost and flexible nature. It can be installed faster than other internet technologies because it can use shorter towers and less cabling, supporting even non-line-of-sight coverage across an entire city or country.

WiMAX isn't just for fixed connections either, like at home. You can also subscribe to a WiMAX service for your mobile devices since USB dongles, laptops, and phones sometimes have the technology built-in.

In addition to internet access, WiMAX can provide voice and video-transferring capabilities as well as telephone access. Since WiMAX transmitters can span a distance of several miles with data rates reaching up to 30-40 megabits per second (1 Gbps for fixed stations), it's easy to see its advantages, especially in areas where wired internet is impossible or too costly to implement.

WiMAX supports several networking usage models:

- A means to transfer data across an Internet Service Provider network — commonly called *backhaul*.

- A form of fixed wireless broadband internet access, replacing satellite internet service.

- A form of mobile internet access that competes directly with LTE technology.

- Internet access for users in extremely remote locations where laying cable would be too expensive.

## WiMAX Disadvantages

Because WiMAX is wireless by nature, the further away from the source that the client gets, the slower their connection becomes. This means that while a user might pull down 30 Mbps in one location, moving away from the cell site can reduce that speed to 1 Mbps or next to nothing.

Similar to when several devices suck away at the bandwidth when connected to a single router, multiple users on one WiMAX radio sector reduce performance for the others.

Wi-Fi is much more popular than WiMAX, so more devices have Wi-Fi capabilities built into them than they do WiMAX. However, most WiMAX implementations include hardware that allows a whole household, for example, to use the service by means of Wi-Fi, much like how a wireless router provides internet for several devices.

## Types of Services Provided by WiMAX

WiMAX can provide two forms of wireless service:

- Non-line-of-sight – Service is a WiFi sort of service. Here a small antenna on your computer connects to the WiMAX tower. In this mode, WiMAX uses a lower frequency range - 2 GHz to 11 GHz (similar to WiFi).

- Line-of-sight – Service, where a fixed dish antenna points straight at the Wi-MAX tower from a rooftop or pole. The line-of-sight connection is stronger and more stable, so it's able to send a lot of data with fewer errors. Line-of-sight transmissions use higher frequencies, with ranges reaching a possible 66 GHz.

## OFDM-based Physical Layer

The WiMAX physical layer (PHY) is based on orthogonal frequency division multiplexing, a scheme that offers good resistance to multipath, and allows WiMAX to operate in NLOS conditions.

## Very High Peak Data Rates

WiMAX is capable of supporting very high peak data rates. In fact, the peak PHY data rate can be as high as 74Mbps when operating using a 20MHz wide spectrum.

More typically, using a 10MHz spectrum operating using TDD scheme with a 3:1 downlink-to-uplink ratio, the peak PHY data rate is about 25Mbps and 6.7Mbps for the downlink and the uplink, respectively.

## Scalable Bandwidth and Data Rate Support

WiMAX has a scalable physical-layer architecture that allows for the data rate to scale easily with available channel bandwidth.

For example, a WiMAX system may use 128, 512, or 1,048-bit FFTs (fast fourier transforms) based on whether the channel bandwidth is 1.25MHz, 5MHz, or 10MHz, respectively. This scaling may be done dynamically to support user roaming across different networks that may have different bandwidth allocations.

## Adaptive Modulation and Coding (AMC)

WiMAX supports a number of modulation and forward error correction (FEC) coding schemes and allows the scheme to be changed as per user and per frame basis, based on channel conditions.

AMC is an effective mechanism to maximize throughput in a time-varying channel.

## Link-layer Retransmissions

WiMAX supports automatic retransmission requests (ARQ) at the link layer for connections that require enhanced reliability. ARQ-enabled connections require each transmitted packet to be acknowledged by the receiver; unacknowledged packets are assumed to be lost and are retransmitted.

## Support for TDD and FDD

IEEE 802.16-2004 and IEEE 802.16e-2005 supports both time division duplexing and frequency division duplexing, as well as a half-duplex FDD, which allows for a low-cost system implementation.

## WiMAX uses OFDM

Mobile WiMAX uses Orthogonal frequency division multiple access (OFDM) as a multiple-access technique, whereby different users can be allocated different subsets of the OFDM tones.

## Flexible and Dynamic per User Resource Allocation

Both uplink and downlink resource allocation are controlled by a scheduler in the base station. Capacity is shared among multiple users on a demand basis, using a burst TDM scheme.

## Support for Advanced Antenna Techniques

The WiMAX solution has a number of hooks built into the physical-layer design, which allows for the use of multiple-antenna techniques, such as beamforming, space-time coding, and spatial multiplexing.

## Quality-of-service Support

The WiMAX MAC layer has a connection-oriented architecture that is designed to support a variety of applications, including voice and multimedia services.

WiMAX system offers support for constant bit rate, variable bit rate, real-time, and non-real-time traffic flows, in addition to best-effort data traffic.

WiMAX MAC is designed to support a large number of users, with multiple connections per terminal, each with its own QoS requirement.

## Robust Security

WiMAX supports strong encryption, using Advanced Encryption Standard (AES), and has a robust privacy and key-management protocol.

The system also offers a very flexible authentication architecture based on Extensible Authentication Protocol (EAP), which allows for a variety of user credentials, including username/password, digital certificates, and smart cards.

### Support for Mobility

The mobile WiMAX variant of the system has mechanisms to support secure seamless handovers for delay-tolerant full-mobility applications, such as VoIP.

### IP-based Architecture

The WiMAX Forum has defined a reference network architecture that is based on an all-IP platform. All end-to-end services are delivered over an IP architecture relying on IP-based protocols for end-to-end transport, QoS, session management, security, and mobility.

## WIMAX ARCHITECTURE

WiMAX technology is based on IEEE standard for high layer protocol such as TCP/IP, VoIP, and SIP etc. WiMAX network is offering air link interoperability and vendor for roaming. The multi vendor access from WiMAX focused on higher level networking specifications for fixed, mobile, and portable WiMAX. The Architecture of WiMAX technology based on all IP platforms. The packet technology of WiMAX needs no legacy circuit telephony. Therefore it reduces the overall cost during life cycle of WiMAX deployment. The main guide lines of WiMAX Architecture are as under.

- The WiMAX architecture support structure of packet switched. WiMAX technology including IEEE 802.16 standard and its modification, suitable for IETF and Ethernet.

- WiMAX architecture allowing decoupling and also sustained topologies for connectivity purpose like IEEE 802.16 radio specifics.

- WiMAX architecture offers flexibility to accommodate a wide range of deployment such as small to large scale. It offers licensed to unlicensed opportunity. WiMAX also support urban, rural radio propagation. The uses of mesh topologies make it more reliable. It is the best co existence of various models.

- WiMAX architecture offers various services and applications such as multimedia, Voice, mandated dogmatic services as emergency and lawful interception.

- WiMAX architecture providing a variety of functions such as ASP, mobile telephony, interface with multi internetworking, media gateway, delivery of IP broadcasting such as MMS, SMS, WAP over IP.

- WiMAX architecture supporting roaming and internetworking. It support wireless network such as 3GPP and 3GPP2.It support wired network as ADSL, MSO based on standard IETF protocols.

- WiMAX architecture also support global roaming, consistent use of AAA for billing purposes, digital certificate, subscriber module, USIM, and RUIM.

- The range of WiMAX architecture is fixed, portable, nomadic, simple mobility and fully mobility.

WiMAX Forum industry represents a logical representation of the WiMAX architecture. The main objective behind WiMAX architecture is to provide amalgamated support needed in a range of network models. The NRM makes out well-designed entities and allusion points accomplished between functional entities. The WiMAX architecture consists of three logical entities MS, ASN, and CSN and reference point for interconnection. All three correspond to a grouping for functional entities which may be single or distributed physical device over several physical devices may be an implementation choice. The manufacturer chooses any implementation according to its choice which is may be individual or combine. The NRM based on the designation of communication protocol and management of data sketch to attain end to end function. It allows manifold implementations for specified useful entity such as mobility and security management.

## Base Station (BS)

The responsibility of Base station (BS) is to provide that the air interface to the MS. The other functionality of BS is micro mobility supervision functions. The handoff prompting, supervision of radio resource, classification of traffic, DHCP, keys, session and multicast group management.

## Access Service Network (ASN)

The ASN (Access Service Network) used to describe an expedient way to explain combination of functional entities and equivalent significance flows connected with the access services. The ASN offers a logical boundary for functional of nearby clients. The connectivity and aggregation services of WiMAX are personified by dissimilar vendors. Planning of functional to logical entities represented in NRM which may execute in unusual ways. The WiMAX forum allows different type of vendors implementation that is interceptive and well-matched for a broad variety of deployment necessities.

## Connectivity Service Network (CSN)

Connectivity Service Network (CSN) is a set of functions related to network offering IP services for connectivity to WiMAX clients. A CSN may include network fundamentals such as AAA, server, routers, and user database and gateway devices that support validation for the devices, services and user. The Connectivity Service Network also handled different type of task such as management of IP addresses, support roaming between different NSPs, management of location, roaming, and mobility between ASNs.

The WiMAX architecture is offering a flexible arrangement of functional entities when constructing the physical entities. Because AS may be molded into BTS, BSC, and an ASNGW. Which are equivalent to the GSM model of BSC, BTS and GPRS Support (SGSN).

# IEEE 802.16

802.16 is a group of broadband wireless communications standards for metropolitan area networks (MANs) developed by a working group of the Institute of Electrical and Electronics Engineers (IEEE). The original 802.16 standard, published in December 2001, specified fixed point-to-multipoint broadband wireless systems operating in the 10-66 GHz licensed spectrum. An amendment, 802.16a, approved in January 2003, specified non-line-of-sight extensions in the 2-11 GHz spectrum, delivering up to 70 Mbps at distances up to 31 miles. Officially called the Wireless MAN specification, 802.16 standards are expected to enable multimedia applications with wireless connection and, with a range of up to 30 miles, provide a viable last mile technology.

An earlier group of IEEE standards, the 802.11 specifications, provide a wireless alternative to Ethernet LANs (local area networks); 802.16 standards are expected to complement these by enabling a wireless alternative to expensive T1 links connecting offices to each other and the Internet. Although the first amendments to the standard are only for fixed wireless connections, a further amendment, 802.16e, is expected to enable connections for mobile devices.

A coalition of wireless industry companies, including Intel, Proxim and Nokia, banded together in April 2001 to form WiMAX, an 802.16 advocacy group. The organization's purpose is to actively promote and certify compatibility and interoperability of devices based on the 802.16 specification, and to develop such devices for the marketplace.

The IEEE 802.16, the Air Interface for Fixed Broadband Wireless Access Systems, also known as the IEEE WirelessMAN air interface, is an emerging suite of standards for fixed, portable and mobile BWA in MAN.

These standards are issued by IEEE 802.16 work group that originally covered the wireless local loop (WLL) technologies in the 10.66 GHz radio spectrum, which were later extended through amendment projects to include both licensed and unlicensed spectra from 2 to 11 GHz.

The WiMAX umbrella currently includes 802.16-2004 and 802.16e. 802.16-2004 utilizes OFDM to serve multiple users in a time division fashion in a sort of a round-robin technique, but done extremely quickly so that users have the perception that they are always transmitting/receiving. 802.16e utilizes OFDMA and can serve multiple users simultaneously by allocating sets of tones to each user.

Table: Following is the chart of various IEEE 802.16 Standards related to WiMAX.

| | 802.16 | 802.16a | 802.16e |
|---|---|---|---|
| Spectrum | 10 - 66 GHz | 2 - 11GHz | <6 GHz |
| Configuration | Line of Sight | Non- Line of Sight | Non- Line of Sight |
| Bit Rate | 32 to 134 Mbps (28 MHz Channel) | ≤ 70 or 100 Mbps (20 MHz Channel) | Up to 15 Mbps |
| Modulation | QPSK,16-QAM, 64-QAM | 256 Sub-Carrier OFDM using QPSK, 16-QAM, 64-QAM 256-QAM | Same as 802.16a |
| Mobility | Fixed | Fixed | ≤75 MPH |
| Channel Bandwidth | 20, 25,28 MHz | Selectable 1.25 to 20 MHz | 5 MHz (Planned) |
| Typical Cell Radius | 1-3 miles | 3-5 miles | 1-3 miles |

The IEEE 802.16 standards for BWA provide the possibility for interoperability between equipment from different vendors, which is in contrast to the previous BWA industry, where proprietary products with high prices are dominant in the market.

# MIMO-WIMAX

The IEEE 802.16e (WiMAX Profile 1.0) and Third Generation Partnership Project (3GPP) Evolved Universal Terrestrial Radio Access (E-UTRA) Long Term Evolution (LTE) (Releases 8 and 9) standards have been developed and are part of the IMT-2000 third generation (3G) technologies. IEEE 802.16m (WiMAX Profile 2.0) and 3GPP E-UTRA LTE-Advanced (LTE-A) (Release 10) are still being developed primarily to meet or exceed the requirements of the International Telecommunication Union (ITU) for IMT Advanced fourth generation (4G) technologies. In this topic we use 802.16m for IEEE 802.16m, LTE for 3GPP releases 8 and 9, LTEA for 3GPP release 10, and E-UTRA for releases 8 to 10. With limited spectrum resources, multiple-input multiple-output (MIMO) techniques are paramount for achieving the minimum target cell spectral efficiency, peak spectral efficiency, and cell edge user spectral efficiency defined by the ITU.

## MIMO Configurations

802.16m and E-UTRA target MIMO schemes for the same sets of antenna configurations: 2, 4, or 8 transmit antennas and a minimum of 2 receive antennas in the DL, and 1, 2, or 4 transmit antennas in the uplink with a minimum of 2 receive antennas. Terminologies in 802.16m and E-UTRA are not matched, and as such could be confusing to the reader. Table gives the equivalence between the terminologies used in both standards. The MIMO systems can be configured as single user MIMO (SU-MIMO), multi-user MIMO (MU-MIMO), and multicell MIMO.

## SU-MIMO

SU-MIMO transmissions occur in time-frequency resources dedicated to a single-terminal advanced mobile station/user equipment (AMS/UE), and allow achieving the peak user spectral efficiency. They encompass techniques ranging from transmit diversity to spatial multiplexing and beamforming. These techniques are supported in 802.16e/m and LTE, with a most noticeable difference in the approach taken for spatial multiplexing.

Spatial multiplexing (SM) is a recognized technique for increasing the peak user throughput by sending multiple spatial streams through multiple antennas, and separating these streams at the receiver by spatial processing. The design of SM techniques in 802.16m and LTE emphasizes trade-offs determined in part by backward compatibility constraints and different assumptions on advanced receiver complexity. While a linear minimum mean square error (LMMSE) receiver is the baseline for performance evaluation, the design should account for the availability of more complex terminals as technology evolves. Design choices for SM also have effects on forward error correction (FEC) encoding coupled with hybrid automatic repeat request (HARQ), feedback mechanisms, and DL control. One of the fundamental choices is the transmission of one or multiple FEC codewords through multiple spatial streams.

Table: Key techniques in MIMO downlink.

| Key downlink MIMO techniques | 802.16m | LTE | LTE-A |
|---|---|---|---|
| Open-loop transmit diversity | SFBC with precoder cycling | SFBC, SFBC+FSTD | Inherited from LTE |
| Open-loop spatial multiplexing | Single codeword with pre- coder cycling | Multiple codewords with large delay CDD | Inherited from LTE |
| Closed-loop spatial multiplexing | Advanced mbeamforming and precoding | Codebook-based precoding, UE- specific RS based beamforming | Advanced beamforming and precoding (under development) |
| Multi-user MIMO | Closed-loop and open-loop MU-MIMO | Closed-loop MU-MIMO | Closed-loop MU-MIMO (under development) |

Table: MIMO capabilities.

| Key down-link MIMO techniques | | 802.16m | 3GPP E-UTRA | | |
|---|---|---|---|---|---|
| | | | LTE | | LTE-A | |
| | | | Release 8 | Release 9 | Release 10 | |
| DL | SU-MIMO | Up to 8 streams | Up to 4 streams | Up to 4 streams | Up to 8 streams |
| | MU-MI-MO | Up to 4 users (non-unitary precoding) | Up to 2 users (unitary precoding) | Up to 4 users (non-unitary precoding)* | Under development |
| UL | SU-MIMO | Up to 4 streams | 1 stream | 1 stream | Up to 4 streams |
| | MU-MI-MO | Up to 4 users | Up to 8 users | Up to 8 users | Under development |
| *Release 8 unitary precoding for up to 2 users is still supported in Releases 9 and 10. | | | | | |

802.16m has evolved from VE transmission adopted in the WiMAX profile Release 1.0. Behind this choice is the assumption that advanced receivers would be better implemented with an optimal two-stream maximum likelihood detector (MLD) than with a successive interference cancellation (MMSE-SIC) detector. This view was continued in 802.16m, relying on promising advances in near MLD techniques such as QRM-MLD or sphere detectors for more than two spatial streams. The usage of VE also facilitates the design and implementation of HARQ processes and requires only a single report of channel quality indicator (CQI) for all the multiplexed layers. Uplink SU-MIMO in 802.16m follows the same design as the DL.

On the other hand, LTE has opted for multiple codewords (MCW) on the DL, while uplink SU-MIMO is still being discussed in LTE-A. MCW allows link adaptation for each FEC codeword in CL SU-MIMO. An MMSESIC receiver may cope with the interference between FEC codewords that are spatially multiplexed. CL SU-MIMO with MCW requires one CQI report and one HARQ process for each FEC codeword. By contrast, layer permutation at the transmitter in OL SU-MIMO with MCW has the effect of averaging the signal-to-noise ratio (SNR) experienced by the two codewords so that a single CQI can be reported, although both codewords still need separate HARQ processes. Note that each FEC codeword in the layer permutation experiences the same channel quality as VE when an LMMSE receiver is used. Therefore, the layer permutation in MCW is equivalent to single codeword (SCW) in some sense. Challenges for accurate modeling of the effective SNR per codeword at the output of an MLD also played a role in the decision to favor MCW with an LMMSE or MMSE-SIC receiver in LTE. Extensive studies during both standardization processes on SCW vs. MCW taken as a part of the whole system under various operation conditions have emphasized the merits of each scheme and led the two standards to develop in their own way.

Table: IEEE 802.16m and 3GPP LTE terminologies.

| 802.16m | E-UTRA |
|---------|--------|
| Advanced base station (ABS) | Enhanced NodeB (eNB) |
| Advanced mobile station (AMS) | User equipment (UE) |
| Transmit antenna | Antenna port |
| Layer | Codeword |
| Stream (i.e., spatial stream) | Layer (i.e., spatial layer) |
| Pilot | Reference signal (RS) |
| Preferred matrix index (PMI) | Precoding matrix indicator (PMI) |
| Dedicated pilots | UE-specific reference signals |
| Vertical encoding (VE) | Single codeword (SCW) |
| Multilayer encoding | Multiple codewords (MCW) |
| Resource unit (RU) | Resource block (RB) |
| Uplink collaborative spatial multiplexing | Uplink MU-MIMO |
| Uplink sounding | Sounding reference signal |

## MU-MIMO

MU-MIMO has become the key technique to fulfill IMT-Advanced requirements. MU-MIMO allocates multiple users in one time-frequency resource to exploit multi-user diversity in the spatial domain, which results in significant gains over SU-MIMO, especially in spatially correlated channels. In configurations such as DL $4 \times 2$ (four transmit antennas and two receive antennas) and uplink $2 \times 4$, single-user transmission only allows spatially multiplexing a maximum of two streams. On the other hand, linear MUMIMO schemes allow sending as many as four spatial streams from four transmit antennas, or receiving as many as four spatial streams with four receive antennas, by multiplexing four spatial streams to or from multiple users. MUMIMO techniques provide large sector throughputs in areas experiencing heavy data traffic.

The different nature of DL demodulation pilots/reference signal (RS) has induced the use of different precoding techniques in 802.16m and E-UTRA. With non-precoded common RSs in LTE, the enhanced NodeB (eNB) needs to signal the index of the precoder to the terminal via a DL control channel. This constrains the precoder to belong to the codebook used to report the precoding matrix indicator or preferred matrix index (PMI). Even though the choice of the actual precoder eventually belongs to the eNB, a simple way of building a precoder is to form a matrix from orthogonal PMIs reported by different users, which leads to unitary precoding.

On the other hand, 802.16m, LTE release 9, and LTE-A have chosen to use DL precoded dedicated pilots (UE-specific RS) even for multiple streams per terminal so that the advanced base station (ABS)/eNB can employ any precoder as long as the same precoder is applied to both RS and data symbols. Both linear and nonlinear MU-MIMO schemes have been considered in the early phase of the two standards. Nonlinear MU-MIMO with dirty paper coding theoretically offers the best performance, but there are many practical limitations to its implementation, so linear MU-MIMO has been adopted in both standards for its simplicity and good performance. Zero-forcing MU-MIMO is one linear MU-MIMO technique commonly assumed in both standards, where the precoding matrix at the transmitter is not unitary. Non-unitary precoding techniques offer significant performance enhancements over unitary precoding, especially in asymmetric antenna configurations (e.g., DL 4 × 2). A scheduler selects several users with good spatial separation and performs pseudo inversion of the combined channel matrix in order to obtain the precoding matrix. The CQI reported by each user is then adjusted at the ABS/eNB to fit the channel quality after precoding. The rank-1 PMI that best approaches the principal eigenvector of the channel matrix and the corresponding CQI need to be reported by the terminal.

When the terminal estimates the rank-1 PMI and CQI, it does not know which other terminal it will be paired with by the scheduler. In 802.16m the AMS estimates the rank-1 PMI and CQI assuming that the paired AMS reports an orthogonal vector. With this approach, inter-user interference is somewhat taken into consideration for MU-MIMO scheduling. From a different perspective, LTE emphasizes the transparent UE operation between SU-MIMO and MUMIMO in terms of CQI feedback, thereby adopting SU-MIMO feedback for MU-MIMO scheduling.

In the DL 802.16m also introduced OL MUMIMO, where a unitary precoding matrix is preset for each frequency-domain resource. Each AMS selects the preferred stream (column of the matrix) for each resource, and reports the corresponding CQI. This technique shows promising performance with limited feedback in uncorrelated and semi-correlated channels. It is suitable for urban areas where user density is high and the channel is typically non-line-of-sight.

For uplink MU-MIMO, both standards allow multiple users to transmit simultaneously in the same uplink resource. The ABS/eNB distinguishes the signals from these terminals through the pilots/RSs allocated to each terminal, and separates them with an advanced receiver, which can be an MLD receiver in the 802.16m orthogonal frequency-division multiple access (OFDMA) uplink, or a turbo MMSE receiver in the LTE single-carrier FDMA uplink.

## Reference Signal for Multi-antenna Operation

An RS, or pilot, is defined in 802.16m and EUTRA to allow measurements of the spatial channel properties and facilitate coherent demodulation at the terminal. An RS can be a dedicated RS (DRS), which is targeted for a specific terminal, or a common RS (CRS), which is shared among a group of terminals. The RS can be further classified into precoded or nonprecoded RS. DRSs are transmitted via virtual antenna ports with a spatial precoding weight to exploit beamforming gain while keeping low RS overhead when the number of virtual antenna ports is smaller than the number of physical antenna ports. CRSs are transmitted via physical antenna ports without a spatial precoder to allow for channel measurements of the non-precoded MIMO channel.

A lot of commonalities exist in 802.16m and E-UTRA in terms of RS usage on the uplink, although exact details are different. For instance, the sounding RS is employed as a non-precoded DRS in both standards for the purpose of uplink spatial channel adaptation, including beam selection and scheduling, as well as for measuring the DL channel by exploiting channel reciprocity in time-division duplex systems. In addition, a precoded DRS is also employed for coherent demodulation in uplink.

In contrast, different designs have been adopted for the DL pilots. Non-precoded common pilots, or midamble, and precoded dedicated pilots are both used in 802.16m for channel measurements and coherent demodulation supporting up to eight transmit antennas. Although the midamble can be used for demodulation in a specific transmission mode such as OL SU-MIMO, it mainly provides fine channel measurements with low pilot overhead because it is transmitted in a wideband manner to enable measurements of the whole frequency band with a small duty cycle. In addition to the midamble, precoded dedicated pilots up to eight streams, which allow flexible beam generation at the ABS side, have been defined for contiguous resource units (RUs) in 802.16m irrespective of the transmission mode. For distributed RUs supporting only two streams, common pilots precoded by predefined matrices are utilized.

On the other hand, non-precoded CRSs supporting up to four antenna ports have been defined for channel measurements and coherent demodulation purposes in LTE Release 8. With the use of non-precoded CRS, the spatial precoder used in the DL signal should be indicated to a terminal in each transmission assignment. LTE Release 8 also supports single-layer beamforming using rank-1 precoded DRS in addition to the CRS, which has been extended to two UE-specific RSs to support dual-layer beamforming in LTE Release 9, and for which the spatial precoder does not need to be indicated to the terminal. To further enhance the peak and average throughputs, eight antenna port transmissions supporting up to eight layers have been adopted in LTE-A. Precoded DRSs supporting up to eight layers and nonprecoded CRSs (i.e., LTE-A channel state information [CSI]-RS) with a low duty cycle for eight antenna ports' measurement are utilized on top of the LTE CRS, which are kept in LTE-A for the continuing support of LTE terminals.

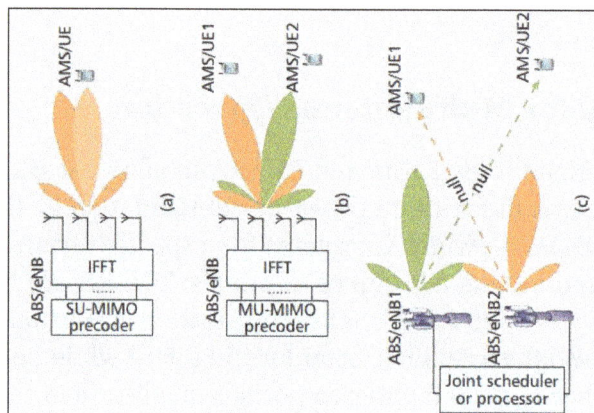

MIMO configurations.

## Open Loop Techniques

In general, OL techniques are designed for high mobility or limited feedback capability. OL techniques mainly provide robustness for the link adaptation considering terminal mobility with infrequent CSI feedback exploiting long-term channel statistics rather than short-term fading information. Therefore, CQI may represent short-term or long-term channel information. As a special case, opportunistic beamforming is also classified as an OL technique since no information relative to the spatial transmit weights is reported, while short-term CQI is used for the adaptation of modulation and coding rate. Two different types of OL techniques have been considered, space-time coding and random beamforming, and these techniques are optimized differently in each standard.

## Space Time/Frequency Code

Transmit diversity techniques provide spatial diversity gain, which translates into higher link margin than single-transmit-antenna techniques. For a configuration with two transmit antennas, both standards have adopted the frequency domain version of the Alamouti code as the basic transmit diversity MIMO mode, where pairs of adjacent subcarriers are coded together instead of two adjacent time slots. Because fast changing channel in the time domain would destroy the orthogonality of the code, an SFBC outperforms an STBC in high-speed scenarios. Application of the SFBC with more than two transmit antennas necessitates the application of a precoder to convert four or eight physical antennas into two virtual antennas. Both techniques adopted in 802.16m and LTE limit the transmission of SFBC to a pair of subcarriers while making effective use of all spatial degrees of freedom over a set of subcarriers in order to provide robustness against spatial correlations in the channel. The receiver can use the same decoding process independent of the number of physical transmit antennas.

CL-MIMO exploits CSI at the transmitter (CSIT) for increasing coverage or throughput. In CL-MIMO, the transmitter acquires the CSI from feedback or channel sounding, and then forms a beamforming or precoding matrix. The major challenges lie in efficiently obtaining the CSI.

The usages of the precoders in 802.16m and LTE are slightly different, due to the difference in the design of DL demodulation pilots, but they target the same objective. While 802.16m chose a combination of precoder cycling and SFBC with precoded pilots, LTE opted for a combination of FSTD and SFBC with non-precoded CRS. The precoder cycling creates a fixed set of two virtual antennas across all subcarriers within an RU and changes the virtual antennas by using different precoder weights across RUs, while FSTD cycles transmissions over pairs of transmit antennas across subcarriers within an RU. These designs also incur different constraints on the channel estimation at the receiver, with a trade-off between the reduced overhead offered by precoded pilots in 802.16m vs. the wider range of interpolation available in the frequency domain with non-precoded pilots for finer channel estimation in 3GPP LTE.

## Random Beamforming/Precoder Cycling

Space-frequency codes exploit spatial diversity so that the variance of the CQI is reduced as the diversity order increases, which allows for robust transmission with infrequent CSI feedback. However, although the robustness of space-frequency codes is superior to other OL schemes, their limited design flexibility led both standards to additionally employ random beamforming. Random beamforming artificially increases the channel selectivity by changing beams within allocated time/frequency resources, and strong FEC codes (e.g., turbo codes) enjoy this artificial frequency diversity gain.

Precoder cycling, in which predefined precoders are cyclically allocated to a group of contiguous subcarriers, is utilized in both standards as a random beamforming technique. The predefined set of precoders is selected from the precoding codebook as a subset, which has a good Chordal distance property so that diversity order can be maximized. Precoder cycling is used for providing beam diversity gain and beam selection gain in 802.16m. To obtain the beam diversity gain, resources are distributed within a wide frequency band where predefined precoders in each localized frequency band form different beams; hence, the aggregated resources at the receiver may enjoy the beam diversity gain. To enable the beam selection gain, on the other hand, a localized resource is allocated to a terminal based on the CQI feedback for its preferred subbands. Since the predefined precoders are cyclically changed according to the subbands, opportunistic beamforming gain can be achieved by allocating the preferred subbands as reported by a terminal. Since non-precoded CRSs are employed in LTE, the predefined precoders can be changed every few subcarriers within each RB so that the beam diversity gains are fully exploited even in a single RB allocation. Layer permutation is performed along with precoder cycling in E-UTRA to further increase diversity gain from virtual antennas with MCW transmissions. The combination of precoder cycling and layer permutation is called large-delay CDD and was adopted as an OL SM technique in EUTRA.

## Closed Loop Techniques

CL-MIMO exploits CSI at the transmitter (CSIT) for increasing coverage or throughput. In CL-MIMO the transmitter acquires the CSI from feedback or channel sounding, and then forms a beamforming or precoding matrix. The major challenges lie in efficiently obtaining the CSI. For accurate CSIT, frequent update is required for mobile terminals. However, overhead and delay limit CSIT accuracy. First, the frame structure inherently sets a delay between the channel measurement and the actual beamforming transmission. Any channel variation during the delay degrades the performance. Second, the overhead for acquiring CSIT becomes burdensome as the number of reporting terminals and mobility increase. On one hand, the CSI of multiple terminals is collected, but only a few favorable terminals are scheduled for transmission. Unfortunately, this selection gain increases logarithmically with the number of reporting terminals. On the other hand, the feedback/sounding overhead increases linearly with the number of reporting terminals. Therefore, efficient feedback/sounding techniques are essential.

## Feedback Mechanisms

Feedback is required when channel reciprocity is unavailable (e.g., in frequency-division duplex systems). The major challenge lies in how to report the preferred beamforming matrix (directions), which is used for the transmitter to compute the actual precoder over a limited feedback bandwidth. For overhead reduction, the whole beamforming matrix is quantized by a matrix or vector codebook. The index of the selected quantization codeword is fed back. An L bit codebook consists of 2L codewords, where L is the required number of bits for indexing each codeword. Here, the term codeword means the quantization codeword in the quantization codebook. In the DL, after measuring the channel, the terminal searches for the best codeword in the codebook for optimizing the performance and reports the PMI to the ABS/eNB. After receiving the PMI, the ABS/eNB looks up the codeword and computes a precoder. Three feedback types are devised, called base codebook, adaptive (or transformed) codebook, and differential codebook, respectively. The base codebook has the least signaling overhead, and the other two have better performance with additional signaling overhead.

The engineering considerations of the base codebook design comprise performance gain, overhead, robustness, complexity, and power amplifier imbalance. First, 802.16m defines 3-bit for 2-transmit antennas (2-Tx) as well as 4-bit and 6-bit feedbacks for 4-transmit antennas (4- Tx), while LTE defines 2-bit and 4-bit feedbacks for 2-Tx and 4-Tx, respectively. Besides the preferred beamforming matrix, an indication of the preferred number of spatial streams is also defined. In addition, 802.16m has 4-bit feedback for 8-transmit antennas (8-Tx) and an enhanced 6-bit feedback for 4-Tx. Second, codewords with a rotated block diagonal structure are explicitly employed in 802.16m and LTE for dual-pole antennas. Third, base codebooks can be dynamically generated from a few parameters for reducing storage complexity. In addition, the codeword entries of all LTE base codebooks and most of the 802.16m codebooks are selected from quaternary phase shift keying (QPSK) and 8-PSK constellations for reducing the storage requirement and computational complexity. Furthermore, the high-rank codewords with more columns include the low rank codewords with a few columns as a subset. This reduces the complexity of searching for the best number of spatial streams. Finally, each LTE codeword and most of the 802.16m codewords load transmission power evenly on each antenna for lowering the power amplifier cost.

Since the optimal codebook varies with the deployment scenario, adaptive codebook is defined in 802.16m. The adaptive codebook changes its codeword distribution according to long-term channel statistics captured in the transmit covariance matrix, which characterizes the spatial correlations across transit-side antennas. Using that matrix, each vector codeword of the rank-1 base codebook is linearly transformed and normalized for generating a codeword in the adaptive codebook. Effectively, more codewords are steered around the directions where the ideal beamforming direction likely appears. As a result, the overall quantization error is reduced. This gain increases with antenna correlation, which increases as the antenna spacing and the angle spread of departing signals at the ABS/eNB decrease. Since MU-MIMO has higher gains in highly correlated channels, the adaptive codebook is most beneficial for MU-MIMO. Antenna configurations, inaccurately calibrated transceiver chains, and propagation channel properties are inherently captured in the measured covariance matrix, making the adaptive codebook robust in a wide range of scenarios. The adaptive codebook, however, requires additional signaling and feedback overhead for reporting the covariance matrix that is needed once for the whole frequency band and for a period greater than 20 ms.

For overhead reduction, the correlation between consecutive beamforming reports is also exploited by differential codebooks in 802.16m. Instead of depicting the preferred beamforming matrices in full, each differential feedback only specifies the incremental change between the current and previous matrices. Because the change is usually within a small range, fewer codewords are needed to cover the small range than the whole beamforming space covered by the base codebook. The reduction of codebook size not only saves feedback overhead but also reduces the quantization complexity. 2-bit, 4-bit, and 4-bit codebooks are defined in 802.16m for 2-Tx, 4-Tx, and 8-Tx, respectively. The down side of differential codebook is the error propagation effect. That is, once an error occurs, that error corrupts the subsequent feedback reports until the differential process is reset.

## Uplink Sounding

In time-division duplex systems, the transmitter can learn about the DL channel from sounding on the uplink channel by exploiting the reciprocity of the propagation channel. To achieve full reciprocity, calibration of the transceiver RF chains is needed at the ABS/eNB. Both 802.16m and LTE provide sounding mechanisms for estimating the uplink channel on a subband or wideband scale. In 802.16m, the uplink sounding channel is inherited from 802.16e with some enhancements. In LTE the ABS/eNB assigns different training sequences to multiple terminals for sharing the same training resources simultaneously. Finally, channel estimation error due to noise and intercell interference limits the performance of uplink sounding.

## Long Term Beamforming

Long term beamforming applies a rough precoder for the whole frequency band and over a period of several frames. This reduces the density of feedback and sounding multiple times. The rough beamforming direction corresponds to a dominant multipath component. Furthermore, in line-of-sight scenarios with closely spaced antennas, the beam pattern is wide in space, and thus allows the beam to cover a high-speed terminal for several time frames. Both 802.16m and LTE support PMI feedback for long-term beamforming. In addition, 802.16m also allows using the reported transmit covariance matrix to derive the precoder.

Long term beamforming applies a rough precoder for the whole frequency band and over a period of several frames. This reduces the density of feedback and sounding by multiple times. The rough beamforming direction corresponds to a dominant multipath component.

# WIBRO

South Korea developed the WiBro (Wireless Broadband) technology in order to provide a high speed Internet solution for local consumers. The technology is based on the 802.16e standard which is more commonly known as the WiMAX standard in other countries. At the time of this writing, the capacity of the WiBro connections available in South Korea have been expanded to 10 Mbit/second which represents a ten-fold increase in capacity compared to earlier WiBro speeds and is a significant upgrade or complement to the 1Gbit/second fiber-optic network that is active in the country.

The WiBro standard was created in order to help overcome the existing data-rate limitations of CDMA mobile phones and to add additional mobility to existing high-speed Internet access services (WiFi and ADSL). The standard uses the OFDMA technology to allow multiple access, an 8.75/10.00 MHz channel bandwidth, and TDD for duplexing. By February of 2002, the South Korean government assigned 100 MHz of the electromagnetic spectrum in the 2.3 to 2.4 GHz band for WiBro use. Towards the end of 2004, the first phase of WiBro was standardized by South Korea and by the end of 2005, the iTU published WiBro under IEEE standard 802.16e (also referred to as mobile WiMAX). In mid-2006, commercial services were released to the general public at a monthly rate of approximately $30 USD equivalent.

## WiBro Network Development

One of the significant steps forward in WiBro and WiMAX technology development occurred in November of 2004 when Samsung and Intel announced an agreement to make sure that WiMAX and WiBro technologies would be compatible with each other. In early 2005, SK Telecom, Hanaro Telecom, and the KT Corporation were chosen to be WiBro operators. In the spring of that year, Hanaro Telecom cancelled their plans to deploy WiBro and returned their license for deploying the network technology.

Several months later, Samsung signed a contract with Sprint Nextel to provide hardware to be used in a WiBro trial run. Two months later, Korea Telecom was able to demonstrate their WiBro services during the APEC (Asia-Pacific Economic Cooperation) summit. During the 2006 Winter Olympics that were held in Turin, Telecom Italia teamed up with Korean Samsung Electronics to publicly demonstrate a WiBro network service. The local service provided download speeds of 10 Mbit/second to users and supported vehicle movement up to speeds of 120 km/hour during this period.

In June of that year, KT Corporation launched their commercial WiBro service in Korea with SK Telecom debuting later than month in Seoul.

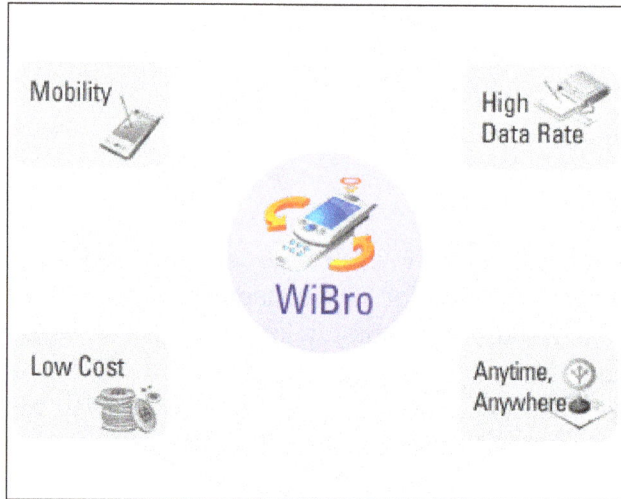

In the spring of 2007, KT announced the launch of their WiBro sercie that included coverage for all subway lines and all areas of Seoul. By early 2011, KT merged their mobile SHOW network and the home network QOOK. By the spring of 2011, the olleh (rebranded KT WiBro network) network had been expanded to include more than 85% of the population and by the end of 2012 close to 90% of all South Koreans.

## Advantages of WiBro

Any given WiBro base station is able to provide an aggregate data rate of 30 to 50 Mbit/second per carrier. These stations see an effective range between one and five kilometers that allows end-users to enjoy high-speed Internet service. The service has also been found to be effective for vehicles moving at speeds of up to approximately 75 mph (miles per hour) and mobile phones seeing mobility speeds up to 150 or more mph.

Although these rates are the max "theoretical" that can be seen by a device, WiBro has also expanded to include Quality of Service (QoS) support. This allows consumers to stream movies, video, and other content that is "loss-sensitive" over the network in a manger that is considered "reliable" by the user. By including QOS support, WiBro is able to support video streaming and related services which provides a major advantage when compared to the base-line WiMAX standard.

Over a similar timeframe, commercialization of WiBro/Mobile WiMAX has also been attempted in several countries. These along with the major service provider(s) include: PORTUS (Croatia), Omnivision (Venezuela), TVA (Brazil), TI (Italy), and Arialink (In Michigan/U.S.). Over time, WiBro has been found to be a bit more restrictive on companies who choose to implement the technology, while WiMAX insists on ensuring interoperability between designs but leaves the implementation detail up to the company.

## Data Speeds of WiBro

At the time of this writing, the maximum rated download and upload speeds of WiBro are:

- Peak Upload Speed: 56 Mbit/s.

- Peak Download Speed: 128 Mbit/s.

## References

- What-is-wimax, wimax: tutorialspoint.com, Retrieved 02 July, 2019

- What-is-wimax-802-16-technology-basics, connectivity-wimax: electronics-notes.com, Retrieved 27 May, 2019

- Wimax-wireless-networking-818321: lifewire.com, Retrieved 19 June, 2019

- Wimax-salient-features, wimax: tutorialspoint.com, Retrieved 10 April, 2019

- Wimax-architecture: freewimaxinfo.com, Retrieved 09 January, 2019

- Wimax-standards, wimax: tutorialspoint.com, Retrieved 25 August, 2019

- MIMO-LTE-WIMAX: iith.ac.in, Retrieved 13 July, 2019

- Wibro: tech-faq.com, Retrieved 23 April, 2019

# CHAPTER 2

# Wireless Network and its Types

Wireless network is a type of network that uses radio signals for communication among computers and other network devices. Wide area network, local area network and personal area network are its three major types. This chapter has been carefully written to provide an easy understanding of the wireless network and its types.

## WIRELESS

Wireless is an encompassing term that describes numerous communication technologies that rely on a wireless signal to send data rather than using a physical medium (often a wire). In wireless transmission, the medium used is the air, through electromagnetic, radio and microwave signals. The term communication here not only means communication between people but between devices and other technologies as well.

Wireless technology started in the early 20th century with radiotelegraphy using Morse code. When the process of modulation was introduced, it became possible to transmit voices, music and other sounds wirelessly. This medium then came to be known as radio. Due to the demand of data communication, the need for a larger portion of the spectrum of wireless signals became a requirement and the term wireless gained widespread use.

When the word wireless is mentioned, people most often mean wireless computer networking as in Wi-Fi or cellular telephony, which is the backbone of personal communications.

Common everyday wireless technologies include:

- 802.11 Wi-Fi: Wireless networking technology for personal computers.

- Bluetooth: Technology for interconnecting small devices.

- Global System for Mobile Communication (GSM): De facto mobile phone standard in many countries.

- Two-Way Radio: Radio communications, as in amateur and citizen band radio services, as well as business and military communications.

# WIRELESS NETWORK

A wireless network is a computer network that uses wireless data connections between network nodes.

Wireless networking is a method by which homes, telecommunications networks and business installations avoid the costly process of introducing cables into a building, or as a connection between various equipment locations. admin telecommunications networks are generally implemented and administered using radio communication. This implementation takes place at the physical level (layer) of the OSI model network structure.

Examples of wireless networks include cell phone networks, wireless local area networks (WLANs), wireless sensor networks, satellite communication networks, and terrestrial microwave networks.

Wireless icon.

## Wireless Links

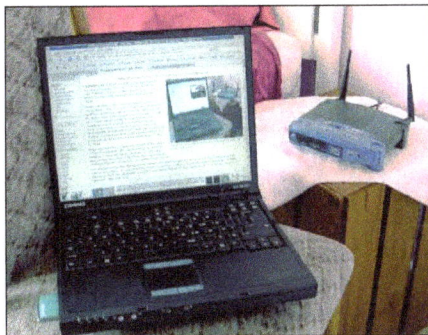

Computers are very often connected to networks using wireless links, e.g. WLANs.

- Terrestrial microwave: Terrestrial microwave communication uses Earth-based transmitters and receivers resembling satellite dishes. Terrestrial microwaves are in the low gigahertz range, which limits all communications to line-of-sight. Relay stations are spaced approximately 48 km (30 mi) apart.

- Communications satellites: Satellites communicate via microwave radio waves, which are not deflected by the Earth's atmosphere. The satellites are stationed in space, typically in geosynchronous orbit 35,400 km (22,000 mi) above the equator. These Earth-orbiting systems are capable of receiving and relaying voice, data, and TV signals.

- Cellular and PCS systems use several radio communications technologies. The systems divide the region covered into multiple geographic areas. Each area has a low-power transmitter or radio relay antenna device to relay calls from one area to the next area.

- Radio and spread spectrum technologies: Wireless local area networks use a high-frequency radio technology similar to digital cellular and a low-frequency radio technology. Wireless LANs use spread spectrum technology to enable communication between multiple devices in a limited area. IEEE 802.11 defines a common flavor of open-standards wireless radio-wave technology.

- Free-space optical communication uses visible or invisible light for communications. In most cases, line-of-sight propagation is used, which limits the physical positioning of communicating devices.

## Types of Wireless Networks

### Wireless PAN

Wireless personal area networks (WPANs) connect devices within a relatively small area, that is generally within a person's reach. For example, both Bluetooth radio and invisible infrared light provides a WPAN for interconnecting a headset to a laptop. ZigBee also supports WPAN applications. Wi-Fi PANs are becoming commonplace as equipment designers start to integrate Wi-Fi into a variety of consumer electronic devices. Intel "My WiFi" and Windows 7 "virtual Wi-Fi" capabilities have made Wi-Fi PANs simpler and easier to set up and configure.

### Wireless LAN

A wireless local area network (WLAN) links two or more devices over a short distance using a wireless distribution method, usually providing a connection through an access point for internet access. The use of spread-spectrum or OFDM technologies may allow users to move around within a local coverage area, and still remain connected to the network.

Products using the IEEE 802.11 WLAN standards are marketed under the Wi-Fi brand name . Fixed wireless technology implements point-to-point links between computers or networks at two distant locations, often using dedicated microwave or modulated laser light beams over line of sight paths. It is often used in cities to connect networks in two or more buildings without installing a wired link. To connect to Wi-Fi, sometimes are used devices like a router or connecting HotSpot using mobile smartphones.

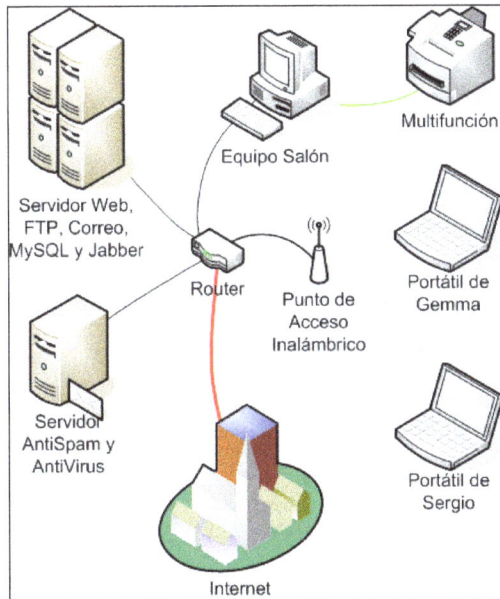

Wireless LANs are often used for connecting to local resources and to the Internet.

## Wireless ad Hoc Network

A wireless ad hoc network, also known as a wireless mesh network or mobile ad hoc network (MANET), is a wireless network made up of radio nodes organized in a mesh topology. Each node forwards messages on behalf of the other nodes and each node performs routing. Ad hoc networks can "self-heal", automatically re-routing around a node that has lost power. Various network layer protocols are needed to realize ad hoc mobile networks, such as Distance Sequenced Distance Vector routing, Associativity-Based Routing, Ad hoc on-demand Distance Vector routing, and Dynamic source routing.

## Wireless MAN

Wireless metropolitan area networks are a type of wireless network that connects several wireless LANs.

WiMAX is a type of Wireless MAN and is described by the IEEE 802.16 standard.

## Wireless WAN

Wireless wide area networks are wireless networks that typically cover large areas, such as between neighbouring towns and cities, or city and suburb. These networks can be used to connect branch offices of business or as a public Internet access system. The wireless connections between access points are usually point to point microwave links using parabolic dishes on the 2.4 GHz and 5.8Ghz band, rather than omnidirectional antennas used with smaller networks. A typical system contains base station gateways, access points and wireless bridging relays. Other configurations are mesh systems where each access point acts as a relay also. When combined with renewable energy systems such as photovoltaic solar panels or wind systems they can be stand alone systems.

## Cellular Network

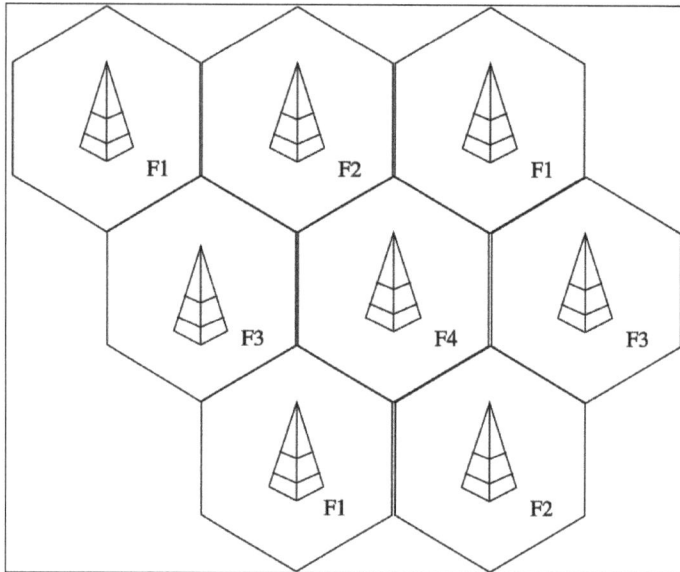

Example of frequency reuse factor or pattern 1/4.

A cellular network or mobile network is a radio network distributed over land areas called cells, each served by at least one fixed-location transceiver, known as a cell site or base station. In a cellular network, each cell characteristically uses a different set of radio frequencies from all their immediate neighbouring cells to avoid any interference.

When joined together these cells provide radio coverage over a wide geographic area. This enables a large number of portable transceivers (e.g., mobile phones, pagers, etc.) to communicate with each other and with fixed transceivers and telephones anywhere in the network, via base stations, even if some of the transceivers are moving through more than one cell during transmission.

Although originally intended for cell phones, with the development of smartphones, cellular telephone networks routinely carry data in addition to telephone conversations:

- Global System for Mobile Communications (GSM): The GSM network is divided into three major systems: the switching system, the base station system, and the operation and support system. The cell phone connects to the base system station which then connects to the operation and support station; it then connects to the switching station where the call is transferred to where it needs to go. GSM is the most common standard and is used for a majority of cell phones.

- Personal Communications Service (PCS): PCS is a radio band that can be used by mobile phones in North America and South Asia. Sprint happened to be the first service to set up a PCS.

- D-AMPS: Digital Advanced Mobile Phone Service, an upgraded version of AMPS, is being phased out due to advancement in technology. The newer GSM networks are replacing the older system.

## Global Area Network

A global area network (GAN) is a network used for supporting mobile across an arbitrary number of wireless LANs, satellite coverage areas, etc. The key challenge in mobile communications is handing off user communications from one local coverage area to the next. In IEEE Project 802, this involves a succession of terrestrial wireless LANs.

## Space Network

Space networks are networks used for communication between spacecraft, usually in the vicinity of the Earth. The example of this is NASA's Space Network.

## Uses

Some examples of usage include cellular phones which are part of everyday wireless networks, allowing easy personal communications. Another example, Intercontinental network systems, use radio satellites to communicate across the world. Emergency services such as the police utilize wireless networks to communicate effectively as well. Individuals and businesses use wireless networks to send and share data rapidly, whether it be in a small office building or across the world.

## Properties

In a general sense, wireless networks offer a vast variety of uses by both business and home users.

"Now, the industry accepts a handful of different wireless technologies. Each wireless technology is defined by a standard that describes unique functions at both the Physical and the Data Link layers of the OSI model. These standards differ in their specified signaling methods, geographic ranges, and frequency usages, among other things. Such differences can make certain technologies better suited to home networks and others better suited to network larger organizations."

## Performance

Each standard varies in geographical range, thus making one standard more ideal than the next depending on what it is one is trying to accomplish with a wireless network. The performance of wireless networks satisfies a variety of applications such as voice and video. The use of this technology also gives room for expansions, such as from 2G to 3G and, 4G and 5G technologies, which stand for the fourth and fifth generation of cell phone mobile communications standards. As wireless networking has become commonplace, sophistication increases through configuration of network hardware and software, and greater capacity to send and receive larger amounts of data, faster, is achieved. Now the wireless network has been running on LTE, which is a 4G mobile communication standard. Users of an LTE network should have data speeds that are 10x faster than a 3G network.

## Space

Space is another characteristic of wireless networking. Wireless networks offer many advantages when it comes to difficult-to-wire areas trying to communicate such as across a street or river, a warehouse on the other side of the premises or buildings that are physically separated but operate as one. Wireless networks allow for users to designate a certain space which the network will be able to communicate with other devices through that network.

Space is also created in homes as a result of eliminating clutters of wiring. This technology allows for an alternative to installing physical network mediums such as TPs, coaxes, or fiber-optics, which can also be expensive.

## Home

For homeowners, wireless technology is an effective option compared to Ethernet for sharing printers, scanners, and high-speed Internet connections. WLANs help save the cost of installation of cable mediums, save time from physical installation, and also creates mobility for devices connected to the network. Wireless networks are simple and require as few as one single wireless access point connected directly to the Internet via a router.

## Wireless Network Elements

The telecommunications network at the physical layer also consists of many interconnected wireline network elements (NEs). These NEs can be stand-alone systems or products that are either supplied by a single manufacturer or are assembled by the service provider (user) or system integrator with parts from several different manufacturers.

Wireless NEs are the products and devices used by a wireless carrier to provide support for the backhaul network as well as a mobile switching center (MSC).

Reliable wireless service depends on the network elements at the physical layer to be protected against all operational environments and applications.

What are especially important are the NEs that are located on the cell tower to the base station (BS) cabinet. The attachment hardware and the positioning of the antenna and associated closures and cables are required to have adequate strength, robustness, corrosion resistance, and resistance against wind, storms, icing, and other weather conditions. Requirements for individual components, such as hardware, cables, connectors, and closures, shall take into consideration the structure to which they are attached.

## Difficulties

## Interference

Compared to wired systems, wireless networks are frequently subject to electromagnetic interference. This can be caused by other networks or other types of equipment that generate radio waves that are within, or close, to the radio bands used for communication. Interference can degrade the signal or cause the system to fail.

## Absorption and Reflection

Some materials cause absorption of electromagnetic waves, preventing it from reaching the receiver, in other cases, particularly with metallic or conductive materials reflection occurs. This can cause dead zones where no reception is available. Aluminium foiled thermal isolation in modern homes can easily reduce indoor mobile signals by 10 dB frequently leading to complaints about the bad reception of long-distance rural cell signals.

## Multipath Fading

In multipath fading two or more different routes taken by the signal, due to reflections, can cause the signal to cancel out at certain locations, and to be stronger in other places (upfade).

## Hidden Node Problem

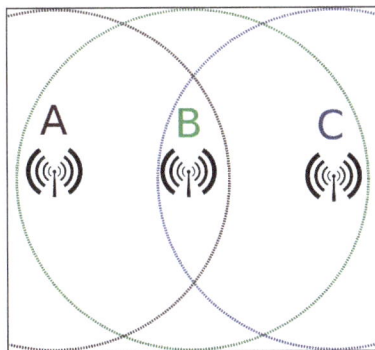

In a hidden node problem Station A can communicate with Station B. Station C can also communicate with Station B. However, Stations A and C cannot communicate with each other, but their signals can interfere at B.

The hidden node problem occurs in some types of network when a node is visible from a wireless access point (AP), but not from other nodes communicating with that AP. This leads to difficulties in media access control (collisions).

## Exposed Terminal Node Problem

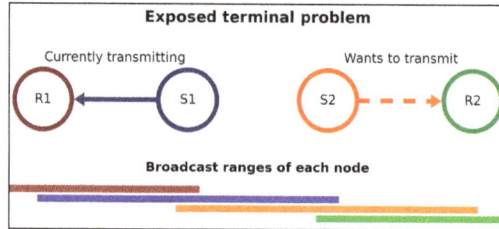

The exposed terminal problem is when a node on one network is unable
to send because of co-channel interference from a node that is on a different network.

## Shared Resource Problem

The wireless spectrum is a limited resource and shared by all nodes in the range of its transmitters. Bandwidth allocation becomes complex with multiple participating users. Often users are not aware that advertised numbers (e.g., for IEEE 802.11 equipment or LTE networks) are not their capacity, but shared with all other users and thus the individual user rate is far lower. With increasing demand, the capacity crunch is more and more likely to happen. User-in-the-loop (UIL) may be an alternative solution to ever upgrading to newer technologies for over-provisioning.

## Capacity

## Channel

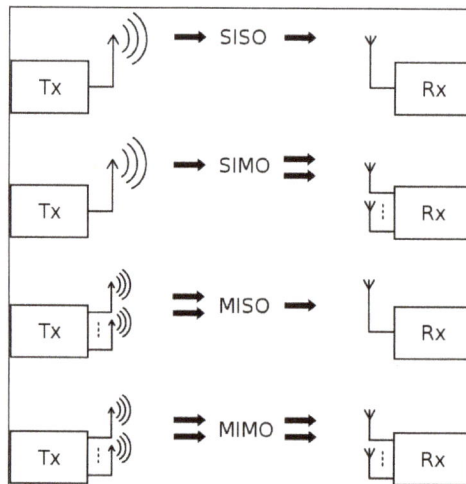

Understanding of SISO, SIMO, MISO and MIMO. Using multiple antennas
and transmitting in different frequency channels can reduce fading,
and can greatly increase the system capacity.

Shannon's theorem can describe the maximum data rate of any single wireless link, which relates to the bandwidth in hertz and to the noise on the channel.

One can greatly increase channel capacity by using MIMO techniques, where multiple aerials or multiple frequencies can exploit multiple paths to the receiver to achieve much higher throughput – by a factor of the product of the frequency and aerial diversity at each end.

Under Linux, the Central Regulatory Domain Agent (CRDA) controls the setting of channels.

## Network

The total network bandwidth depends on how dispersive the medium is (more dispersive medium generally has better total bandwidth because it minimises interference), how many frequencies are available, how noisy those frequencies are, how many aerials are used and whether a directional antenna is in use, whether nodes employ power control and so on.

Cellular wireless networks generally have good capacity, due to their use of directional aerials, and their ability to reuse radio channels in non-adjacent cells. Additionally, cells can be made very small using low power transmitters this is used in cities to give network capacity that scales linearly with population density.

## Safety

Wireless access points are also often close to humans, but the drop off in power over distance is fast, following the inverse-square law. The position of the United Kingdom's Health Protection Agency (HPA) is that "radio frequency (RF) exposures from WiFi are likely to be lower than those from mobile phones." It also saw "no reason why schools and others should not use WiFi equipment." In October 2007, the HPA launched a new "systematic" study into the effects of WiFi networks on behalf of the UK government, in order to calm fears that had appeared in the media in a recent period up to that time". Dr Michael Clark, of the HPA, says published research on mobile phones and masts does not add up to an indictment of WiFi.

## WIRELESS MESH NETWORK

A wireless mesh network (WMN) is a communications network made up of radio nodes organized in a mesh topology. It is also a form of wireless ad hoc network.

A mesh refers to rich interconnection among devices or nodes. Wireless mesh networks often consist of mesh clients, mesh routers and gateways. Mobility of nodes is less frequent.

If nodes constantly or frequently move, the mesh spends more time updating routes than delivering data. In a wireless mesh network, topology tends to be more static, so that routes computation can converge and delivery of data to their destinations can occur. Hence, this is a low-mobility centralized form of wireless ad hoc network. Also, because it sometimes relies on static nodes to act as gateways, it is not a truly all-wireless ad hoc network.

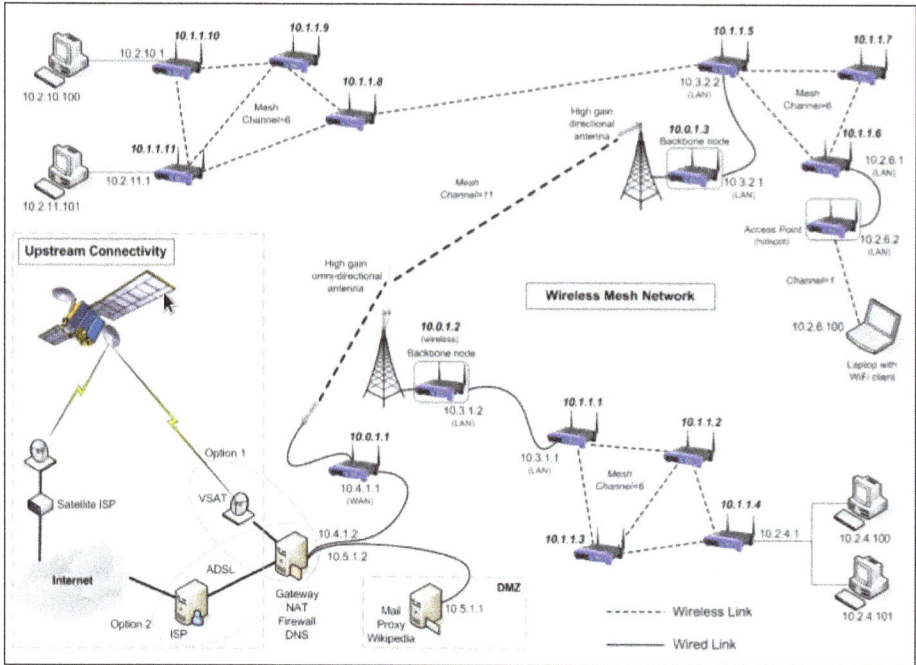

Diagram showing a possible configuration for a wired-wireless mesh
network, connected upstream via a VSAT link.

Mesh clients are often laptops, cell phones, and other wireless devices. Mesh routers forward traffic to and from the gateways, which may, but need not, be connected to the Internet. The coverage area of all radio nodes working as a single network is sometimes called a mesh cloud. Access to this mesh cloud depends on the radio nodes working together to create a radio network. A mesh network is reliable and offers redundancy. When one node can no longer operate, the rest of the nodes can still communicate with each other, directly or through one or more intermediate nodes. Wireless mesh networks can self form and self heal. Wireless mesh networks work with different wireless technologies including 802.11, 802.15, 802.16, cellular technologies and need not be restricted to any one technology or protocol.

## Features

## Architecture

Wireless mesh architecture is a first step towards providing cost effective and low mobility over a specific coverage area. Wireless mesh infrastructure is, in effect, a network of routers minus the cabling between nodes. It is built of peer radio devices that do not have to be cabled to a wired port like traditional WLAN access points (AP) do. Mesh infrastructure carries data over large distances by splitting the distance into a series of short hops. Intermediate nodes not only boost the signal, but cooperatively pass data from point A to point B by making forwarding decisions based on their knowledge of the network, i.e. perform routing by first deriving the topology of the network.

Wireless mesh networks is a relatively "stable-topology" network except for the occasional failure of nodes or addition of new nodes. The path of traffic, being aggregated from a large number of end users, changes infrequently. Practically all the traffic in an infrastructure mesh network is either forwarded to or from a gateway, while in wireless ad hoc networks or client mesh networks the traffic flows between arbitrary pairs of nodes.

If rate of mobility among nodes are high, i.e., link breaks happen frequently, wireless mesh networks start to break down and have low communication performance.

## Management

This type of infrastructure can be decentralized (with no central server) or centrally managed (with a central server). Both are relatively inexpensive, and can be very reliable and resilient, as each node needs only transmit as far as the next node. Nodes act as routers to transmit data from nearby nodes to peers that are too far away to reach in a single hop, resulting in a network that can span larger distances. The topology of a mesh network must be relatively stable, i.e., not too much mobility. If one node drops out of the network, due to hardware failure or any other reason, its neighbors can quickly find another route using a routing protocol.

## Applications

Mesh networks may involve either fixed or mobile devices. The solutions are as diverse as communication needs, for example in difficult environments such as emergency situations, tunnels, oil rigs, battlefield surveillance, high-speed mobile-video applications on board public transport, real-time racing-car telemetry, or self-organizing Internet access for communities. An important possible application for wireless mesh networks is VoIP. By using a quality of service scheme, the wireless mesh may support routing local telephone calls through the mesh. Most applications in wireless mesh networks are similar to those in wireless ad hoc networks.

Some current applications:

- U.S. military forces are now using wireless mesh networking to connect their computers, mainly ruggedized laptops, in field operations.

- Electric smart meters now being deployed on residences, transfer their readings from one to another and eventually to the central office for billing, without the need for human meter readers or the need to connect the meters with cables.

- The laptops in the One Laptop per Child program use wireless mesh networking to enable students to exchange files and get on the Internet even though they lack wired or cell phone or other physical connections in their area.

- Google Home, Google Wi-Fi, and Google OnHub all support Wi-Fi mesh (i.e., Wi-Fi ad hoc) networking. Several manufacturers of Wi-Fi routers began offering mesh routers for home use in the mid-2010s.

- The 66-satellite Iridium constellation operates as a mesh network, with wireless links between adjacent satellites. Calls between two satellite phones are routed through the mesh, from one satellite to another across the constellation, without having to go through an earth station. This makes for a smaller travel distance for the signal, reducing latency, and also allows for the constellation to operate with far fewer earth stations than would be required for 66 traditional communications satellites.

## Operation

The principle is similar to the way packets travel around the wired Internet—data hops from one device to another until it eventually reaches its destination. Dynamic routing algorithms implemented in each device allow this to happen. To implement such dynamic routing protocols, each device needs to communicate routing information to other devices in the network. Each device then determines what to do with the data it receives – either pass it on to the next device or keep it, depending on the protocol. The routing algorithm used should attempt to always ensure that the data takes the most appropriate (fastest) route to its destination.

## Multi-radio Mesh

Multi-radio mesh refers to having different radios operating at different frequencies to interconnect nodes in a mesh. This means there is a unique frequency used for each wireless hop and thus a dedicated CSMA collision domain. With more radio bands, communication throughput is likely to increase as a result of more available communication channels. This is similar to providing dual or multiple radio paths to transmit and receive data.

## Protocols

## Routing Protocols

There are more than 70 competing schemes for routing packets across mesh networks. Some of these include:

- Associativity-Based Routing (ABR).

- AODV (Ad hoc On-Demand Distance Vector).

- B.A.T.M.A.N. (Better Approach To Mobile Adhoc Networking).
- Babel (protocol) (a distance-vector routing protocol for IPv6 and IPv4 with fast convergence properties).
- Dynamic NIx-Vector Routing|DNVR.
- DSDV (Destination-Sequenced Distance-Vector Routing).
- DSR (Dynamic Source Routing).
- HSLS (Hazy-Sighted Link State).
- HWMP (Hybrid Wireless Mesh Protocol, the default mandatory routing protocol of IEEE 802.11s).
- Infrastructure Wireless Mesh Protocol (IWMP) for Infrastructure Mesh Networks by GRECO UFPB-Brazil.
- OLSR (Optimized Link State Routing protocol).
- OORP (OrderOne Routing Protocol) (OrderOne Networks Routing Protocol).
- OSPF (Open Shortest Path First Routing).
- Routing Protocol for Low-Power and Lossy Networks (IETF ROLL RPL protocol, RFC 6550).
- PWRP (Predictive Wireless Routing Protocol).
- TORA (Temporally-Ordered Routing Algorithm).
- ZRP (Zone Routing Protocol).

### Autoconfiguration Protocols

Standard autoconfiguration protocols, such as DHCP or IPv6 stateless autoconfiguration may be used over mesh networks.

Mesh network specific autoconfiguration protocols include:

- Ad Hoc Configuration Protocol (AHCP).
- Proactive Autoconfiguration (Proactive Autoconfiguration Protocol).
- Dynamic WMN Configuration Protocol (DWCP).

# WIRELESS SENSOR NETWORK

Wireless sensor network (WSN) refers to a group of spatially dispersed and dedicated sensors for monitoring and recording the physical conditions of the environment and organizing the collected data at a central location. WSNs measure environmental conditions like temperature, sound, pollution levels, humidity, wind, and so on.

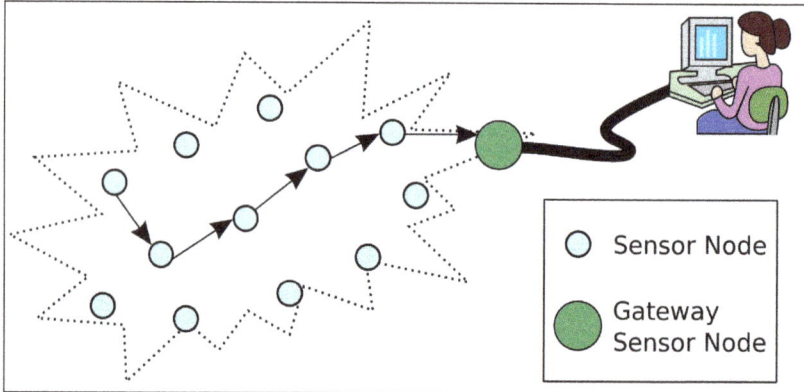

Representation of a wireless sensor network.

These are similar to wireless ad hoc networks in the sense that they rely on wireless connectivity and spontaneous formation of networks so that sensor data can be transported wirelessly. WSNs are spatially distributed autonomous sensors to *monitor* physical or environmental conditions, such as temperature, sound, pressure, etc. and to cooperatively pass their data through the network to a main location. The more modern networks are bi-directional, also enabling *control* of sensor activity. The development of wireless sensor networks was motivated by military applications such as battlefield surveillance; today such networks are used in many industrial and consumer applications, such as industrial process monitoring and control, machine health monitoring, and so on.

The WSN is built of "nodes" – from a few to several hundreds or even thousands, where each node is connected to one (or sometimes several) sensors. Each such sensor network node has typically several parts: a radio transceiver with an internal antenna or connection to an external antenna, a microcontroller, an electronic circuit for interfacing with the sensors and an energy source, usually a battery or an embedded form of energy harvesting. A sensor node might vary in size from that of a shoebox down to the size of a grain of dust, although functioning "motes" of genuine microscopic dimensions have yet to be created. The cost of sensor nodes is similarly variable, ranging from a few to hundreds of dollars, depending on the complexity of the individual sensor nodes. Size and cost constraints on sensor nodes result in corresponding constraints on resources such as energy, memory, computational speed and communications bandwidth. The topology of the WSNs can vary from a simple star network to an advanced multi-hop wireless mesh network. The propagation technique between the hops of the network can be routing or flooding.

In computer science and telecommunications, wireless sensor networks are an active research area with numerous workshops and conferences arranged each year, for example IPSN, SenSys, and EWSN. As of 2010, wireless sensor networks have reached approximately 120 million remote units worldwide.

# Application

## Area Monitoring

Area monitoring is a common application of WSNs. In area monitoring, the WSN is deployed over a region where some phenomenon is to be monitored. A military example is the use of sensors to detect enemy intrusion; a civilian example is the geo-fencing of gas or oil pipelines.

## Health Care Monitoring

There are several types of sensor networks for medical applications: implanted, wearable, and environment-embedded. Implantable medical devices are those that are inserted inside the human body. Wearable devices are used on the body surface of a human or just at close proximity of the user. Environment-embedded systems employ sensors contained in the environment. Possible applications include body position measurement, location of persons, overall monitoring of ill patients in hospitals and at home. Devices embedded in the environment track the physical state of a person for continuous health diagnosis, using as input the data from a network of depth cameras, a sensing floor, or other similar devices. Body-area networks can collect information about an individual's health, fitness, and energy expenditure. In health care applications the privacy and authenticity of user data has prime importance. Especially due to the integration of sensor networks, with IoT, the user authentication becomes more challenging; however, a solution is presented in recent work.

## Environmental/Earth Sensing

There are many applications in monitoring environmental parameters, examples of which are given below. They share the extra challenges of harsh environments and reduced power supply.

## Air Pollution Monitoring

Wireless sensor networks have been deployed in several cities (Stockholm, London, and Brisbane) to monitor the concentration of dangerous gases for citizens. These can take advantage of the ad hoc wireless links rather than wired installations, which also make them more mobile for testing readings in different areas.

## Forest Fire detection

A network of Sensor Nodes can be installed in a forest to detect when a fire has started. The nodes can be equipped with sensors to measure temperature, humidity and gases which are produced by fire in the trees or vegetation. The early detection is crucial for a successful action of the firefighters; thanks to Wireless Sensor Networks, the fire brigade will be able to know when a fire is started and how it is spreading.

## Landslide Detection

A landslide detection system makes use of a wireless sensor network to detect the slight movements of soil and changes in various parameters that may occur before or during a landslide. Through the data gathered it may be possible to know the impending occurrence of landslides long before it actually happens.

## Water Quality Monitoring

Water quality monitoring involves analyzing water properties in dams, rivers, lakes and oceans, as well as underground water reserves. The use of many wireless distributed sensors enables the creation of a more accurate map of the water status, and allows the permanent deployment of monitoring stations in locations of difficult access, without the need of manual data retrieval.

## Natural Disaster Prevention

Wireless sensor networks can be effective in preventing adverse consequences of natural disasters, like floods. Wireless nodes have been deployed successfully in rivers, where changes in water levels must be monitored in real time.

## Industrial Monitoring

## Machine Health Monitoring

Wireless sensor networks have been developed for machinery condition-based maintenance (CBM) as they offer significant cost savings and enable new functionality.

Wireless sensors can be placed in locations difficult or impossible to reach with a wired system, such as rotating machinery and untethered vehicles.

## Data Logging

Wireless sensor networks also are used for the collection of data for monitoring of environmental information. This can be as simple as monitoring the temperature in a fridge or the level of water in overflow tanks in nuclear power plants. The statistical information can then be used to show how systems have been working. The advantage of WSNs over conventional loggers is the "live" data feed that is possible.

## Water/Waste Water Monitoring

Monitoring the quality and level of water includes many activities such as checking the quality of underground or surface water and ensuring a country's water infrastructure for the benefit of both human and animal. It may be used to protect the wastage of water.

## Structural Health Monitoring

Wireless sensor networks can be used to monitor the condition of civil infrastructure and related geo-physical processes close to real time, and over long periods through data logging, using appropriately interfaced sensors.

## Wine Production

Wireless sensor networks are used to monitor wine production, both in the field and the cellar.

## Threat Detection

The Wide Area Tracking System (WATS) is a prototype network for detecting a ground-based nuclear device such as a nuclear "briefcase bomb." WATS is being developed at the Lawrence Livermore National Laboratory (LLNL). WATS would be made up of wireless gamma and neutron sensors connected through a communications network. Data picked up by the sensors undergoes "data fusion", which converts the information into easily interpreted forms; this data fusion is the most important aspect of the system.

The data fusion process occurs *within* the sensor network rather than at a centralized computer and is performed by a specially developed algorithm based on Bayesian statistics. WATS would not use a centralized computer for analysis because researchers found that factors such as latency and available bandwidth tended to create significant bottlenecks. Data processed in the field by the network itself (by transferring small amounts of data between neighboring sensors) is faster and makes the network more scalable.

An important factor in WATS development is *ease of deployment*, since more sensors both improves the detection rate and reduces false alarms. WATS sensors could be deployed in permanent positions or mounted in vehicles for mobile protection of specific locations. One barrier to the implementation of WATS is the size, weight, energy requirements and cost of currently available wireless sensors. The development of improved sensors is a major component of current research at the Nonproliferation, Arms Control, and International Security (NAI) Directorate at LLNL.

WATS was profiled to the U.S. House of Representatives' Military Research and Development Subcommittee on October 1, 1997 during a hearing on nuclear terrorism and countermeasures. On August 4, 1998 in a subsequent meeting of that subcommittee, Chairman Curt Weldon stated that research funding for WATS had been cut by the Clinton administration to a subsistence level and that the program had been poorly re-organized.

## Characteristics

The main characteristics of a WSN include:

- Power consumption constraints for nodes using batteries or energy harvesting. Examples of suppliers are ReVibe Energy and Perpetuum.

- Ability to cope with node failures (resilience).

- Some mobility of nodes.

- Heterogeneity of nodes.

- Homogeneity of nodes.

- Scalability to large scale of deployment.

- Ability to withstand harsh environmental conditions.

- Ease of use.

- Cross-layer optimization.

Cross-layer is becoming an important studying area for wireless communications. In addition, the traditional layered approach presents three main problems:

- Traditional layered approach cannot share different information among different layers, which leads to each layer not having complete information. The traditional layered approach cannot guarantee the optimization of the entire network.

- The traditional layered approach does not have the ability to adapt to the environmental change.

- Because of the interference between the different users, access conflicts, fading, and the change of environment in the wireless sensor networks, traditional layered approach for wired networks is not applicable to wireless networks.

So the cross-layer can be used to make the optimal modulation to improve the transmission performance, such as data rate, energy efficiency, QoS (Quality of Service), etc. Sensor nodes can be imagined as small computers which are extremely basic in terms of their interfaces and their components. They usually consist of a processing unit with limited computational power and limited memory, sensors or MEMS (including specific conditioning circuitry), a communication device (usually radio transceivers or alternatively optical), and a power source usually in the form of a battery. Other possible inclusions are energy harvesting modules, secondary ASICs, and possibly secondary communication interface (e.g. RS-232 or USB).

The base stations are one or more components of the WSN with much more computational, energy and communication resources. They act as a gateway between sensor nodes and the end user as they typically forward data from the WSN on to a server. Other special components in routing based networks are routers, designed to compute, calculate and distribute the routing tables.

## Platforms

### Hardware

One major challenge in a WSN is to produce *low cost* and *tiny* sensor nodes. There are an increasing number of small companies producing WSN hardware and the commercial situation can be compared to home computing in the 1970s. Many of the nodes are still in the research and development stage, particularly their software. Also inherent to sensor network adoption is the use of very low power methods for radio communication and data acquisition.

In many applications, a WSN communicates with a Local Area Network or Wide Area Network through a gateway. The Gateway acts as a bridge between the WSN and the other network. This enables data to be stored and processed by devices with more resources, for example, in a remotely located server. A wireless wide area network used primarily for low-power devices is known as a Low-Power Wide-Area Network (LPWAN).

### Wireless

There are several wireless standards and solutions for sensor node connectivity. Thread and ZigBee can connect sensors operating at 2.4 GHz with a data rate of 250kbit/s. Many use a lower frequency to increase radio range (typically 1 km), for example Z-wave operates at 915 MHz and in the EU 868 MHz has been widely used but these have a lower data rate (typically 50 kb/s). The IEEE 802.15.4 working group provides a standard for low power device connectivity and commonly sensors and smart meters use one of these standards for connectivity. With the emergence of Internet of Things, many other proposals have been made to provide sensor connectivity. LORA is a form of LPWAN which provides long range low power wireless connectivity for devices, which has been used in smart meters. Wi-SUN connects devices at home. NarrowBand IOT and LTE-M can connect up to millions of sensors and devices using cellular technology.

### Software

Energy is the scarcest resource of WSN nodes, and it determines the lifetime of WSNs. WSNs may be deployed in large numbers in various environments, including remote and hostile regions, where ad hoc communications are a key component. For this reason, algorithms and protocols need to address the following issues:

- Increased lifespan,

- Robustness and fault tolerance,

- Self-configuration.

Lifetime maximization: Energy/Power Consumption of the sensing device should be minimized and sensor nodes should be energy efficient since their limited energy resource determines their lifetime. To conserve power, wireless sensor nodes normally power off both the radio transmitter and the radio receiver when not in use.

## Routing Protocols

Wireless sensor networks are composed of low-energy, small-size, and low-range unattended sensor nodes. Recently, it has been observed that by periodically turning on and off the sensing and communication capabilities of sensor nodes, we can significantly reduce the active time and thus prolong network lifetime. However, this duty cycling may result in high network latency, routing overhead, and neighbor discovery delays due to asynchronous sleep and wake-up scheduling. These limitations call for a countermeasure for duty-cycled wireless sensor networks which should minimize routing information, routing traffic load, and energy consumption. Researchers from Sungkyunkwan University have proposed a lightweight non-increasing delivery-latency interval routing referred as LNDIR. This scheme can discover minimum latency routes at each non-increasing delivery-latency interval instead of each time slot. Simulation experiments demonstrated the validity of this novel approach in minimizing routing information stored at each sensor. Furthermore, this novel routing can also guarantee the minimum delivery latency from each source to the sink. Performance improvements of up to 12-fold and 11-fold are observed in terms of routing traffic load reduction and energy efficiency, respectively, as compared to existing schemes.

## Operating Systems

Operating systems for wireless sensor network nodes are typically less complex than general-purpose operating systems. They more strongly resemble embedded systems, for two reasons. First, wireless sensor networks are typically deployed with a particular application in mind, rather than as a general platform. Second, a need for low costs and low power leads most wireless sensor nodes to have low-power microcontrollers ensuring that mechanisms such as virtual memory are either unnecessary or too expensive to implement.

It is therefore possible to use embedded operating systems such as eCos or uC/OS for sensor networks. However, such operating systems are often designed with real-time properties.

TinyOS is perhaps the first operating system specifically designed for wireless sensor networks. TinyOS is based on an event-driven programming model instead of multithreading. TinyOS programs are composed of event handlers and tasks with run-to-completion semantics. When an external event occurs, such as an incoming data packet or a sensor reading, TinyOS signals the appropriate event handler to handle the event. Event handlers can post tasks that are scheduled by the TinyOS kernel some time later.

LiteOS is a newly developed OS for wireless sensor networks, which provides UNIX-like abstraction and support for the C programming language.

Contiki is an OS which uses a simpler programming style in C while providing advances such as 6LoWPAN and Protothreads.

RIOT (operating system) is a more recent real-time OS including similar functionality to Contiki.

PreonVM is an OS for wireless sensor networks, which provides 6LoWPAN based on Contiki and support for the Java programming language.

## Online Collaborative Sensor Data Management Platforms

Online collaborative sensor data management platforms are on-line database services that allow sensor owners to register and connect their devices to feed data into an online database for storage and also allow developers to connect to the database and build their own applications based on that data. Examples include Xively and the Wikisensing platform. Such platforms simplify online collaboration between users over diverse data sets ranging from energy and environment data to that collected from transport services. Other services include allowing developers to embed re-al-time graphs & widgets in websites; analyse and process historical data pulled from the data feeds; send real-time alerts from any datastream to control scripts, devices and environments.

The architecture of the Wikisensing system describes the key components of such systems to include APIs and interfaces for online collaborators, a middleware containing the business logic needed for the sensor data management and processing and a storage model suitable for the efficient storage and retrieval of large volumes of data.

## Simulation

At present, agent-based modeling and simulation is the only paradigm which allows the simulation of complex behavior in the environments of wireless sensors (such as flocking). Agent-based simulation of wireless sensor and ad hoc networks is a relatively new paradigm. Agent-based modelling was originally based on social simulation.

Network simulators like Opnet, Tetcos NetSim and NS can be used to simulate a wireless sensor network.

## Security

Infrastructure-less architecture (i.e. no gateways are included, etc.) and inherent requirements (i.e. unattended working environment, etc.) of WSNs might pose several weak points that attract adversaries. Therefore, security is a big concern when WSNs are deployed for special applications such as military and healthcare. Owing to their unique characteristics, traditional security methods of computer networks would be useless (or less effective) for WSNs. Hence, lack of security mechanisms would cause intrusions towards those networks. These intrusions need to be detected and mitigation methods should be applied.

## Distributed Sensor Network

If a centralized architecture is used in a sensor network and the central node fails, then the entire network will collapse, however the reliability of the sensor network can be increased by using a distributed control architecture. Distributed control is used in WSNs for the following reasons:

- Sensor nodes are prone to failure,

- For better collection of data,

- To provide nodes with backup in case of failure of the central node.

There is also no centralised body to allocate the resources and they have to be self organized.

As for the distributed filtering over distributed sensor network. the general setup is to observe the underlying process through a group of sensors organized according to a given network topology, which renders the individual observer estimates the system state based not only on its own measurement but also on its neighbors'.

## Data Integration and Sensor Web

The data gathered from wireless sensor networks is usually saved in the form of numerical data in a central base station. Additionally, the Open Geospatial Consortium (OGC) is specifying standards for interoperability interfaces and metadata encodings that enable real time integration of heterogeneous sensor webs into the Internet, allowing any individual to monitor or control wireless sensor networks through a web browser.

## In-network Processing

To reduce communication costs some algorithms remove or reduce nodes' redundant sensor information and avoid forwarding data that is of no use. This technique has been used, for instance, for distributed anomaly detection or distributed optimization. As nodes can inspect the data they forward, they can measure averages or directionality for example of readings from other nodes. For example, in sensing and monitoring applications, it is generally the case that neighboring sensor nodes monitoring an environmental feature typically register similar values. This kind of data redundancy due to the spatial correlation between sensor observations inspires techniques for in-network data aggregation and mining. Aggregation reduces the amount of network traffic which helps to reduce energy consumption on sensor nodes. Recently, it has been found that network gateways also play an important role in improving energy efficiency of sensor nodes by scheduling more resources for the nodes with more critical energy efficiency need and advanced energy efficient scheduling algorithms need to be implemented at network gateways for the improvement of the overall network energy efficiency.

## Secure Data Aggregation

This is a form of in-network processing where sensor nodes are assumed to be unsecured with limited available energy, while the base station is assumed to be secure with unlimited available energy. Aggregation complicates the already existing security challenges for wireless sensor networks and requires new security techniques tailored specifically for this scenario. Providing security to aggregate data in wireless sensor networks is known as secure data aggregation in WSN.

Two main security challenges in secure data aggregation are confidentiality and integrity of data. While encryption is traditionally used to provide end to end confidentiality in wireless sensor network, the aggregators in a secure data aggregation scenario need to decrypt the encrypted data to perform aggregation. This exposes the plaintext at the aggregators, making the data vulnerable to attacks from an adversary. Similarly an aggregator can inject false data into the aggregate and make the base station accept false data. Thus, while data aggregation improves energy efficiency of a network, it complicates the existing security challenges.

# WIRELESS APPLICATION SERVICE PROVIDER

A wireless application service provider (WASP) is part of a growing industry sector resulting from the convergence of two trends: wireless communications and the outsourcing of services. A WASP performs the same service for wireless clients as a regular application service provider (ASP) does for wired clients: it provides Web-based access to applications and services that would otherwise have to be stored locally. The main difference with WASP is that it enables customers to access the service from a variety of wireless devices, such as a smartphone or personal digital assistant (PDA).

Although the business world is increasingly mobile, many corporations are resisting the idea of wireless communication, because of concerns about set-up and maintenance costs and the need for in-house expertise. WASPs offer businesses the advantages of wireless service with less expense and fewer risks. Because mobile applications are subscribed to, rather than purchased, up-front costs are lower; because the WASP provides support, staffing and training costs are lower.

WASP services may include:

- Constant system monitoring.

- Diagnostics and resolution.

- User support.

- Text formatting for various devices.

- Problem detection and reporting.

There are still issues to be resolved. Coverage areas remain limited, for example, and data synchronization among devices can be problematic. Nevertheless, WASPs provide an easier, safer, and cheaper way for organizations to add mobile components, and a number of major companies are opting for them. UPS, Sprint, and eBay are among the early subscribers to WASP services. Interestingly, some ASPs have begun to offer WASP services, while others are purchasing them.

# WIRELESS BROADBAND

Wireless broadband is telecommunications technology that provides high-speed wireless Internet access or computer networking access over a wide area. The term comprises both fixed and mobile broadband.

Three fixed wireless dishes with protective covers on top.

## Broadband

Originally the word "broadband" had a technical meaning, but became a marketing term for any kind of relatively high-speed computer network or Internet access technology. According to the 802.16-2004 standard, broadband means "having instantaneous bandwidths greater than 1 MHz and supporting data rates greater than about 1.5 Mbit/s." The Federal Communications Commission (FCC) recently re-defined the definition to mean download speeds of at least 25 Mbit/s and upload speeds of at least 3 Mbit/s.

## Technology and Speeds

A typical WISP Customer Premises Equipment (CPE) installed on a residence.

Wireless networks can feature data rates roughly equivalent to some wired networks, such as that of asymmetric digital subscriber line (ADSL) or a cable modem. Wireless networks can also be symmetrical, meaning the same rate in both directions (downstream and upstream), which is most commonly associated with fixed wireless networks. A fixed wireless network link is a stationary terrestrial wireless connection, which can support higher data rates for the same power as mobile or satellite systems.

Few wireless Internet service providers (WISPs) provide download speeds of over 100 Mbit/s; most broadband wireless access (BWA) services are estimated to have a range of 50 km (31 mi) from a tower. Technologies used include Local Multipoint Distribution Service (LMDS) and Multichannel Multipoint Distribution Service (MMDS), as well as heavy use of the industrial, scientific and medical (ISM) radio bands and one particular access technology was standardized by IEEE 802.16, with products known as WiMAX.

WiMAX is highly popular in Europe but has not met full acceptance in the United States because cost of deployment does not meet return on investment figures. In 2005 the Federal Communications Commission adopted a Report and Order that revised the FCC's rules to open the 3650 MHz band for terrestrial wireless broadband operations.

## Development of Wireless Broadband

On November 14, 2007 the Commission released Public Notice DA 07-4605 in which the Wireless Telecommunications Bureau announced the start date for licensing and registration process for the 3650–3700 MHz band. In 2010 the FCC adopted the TV White Space Rules (TVWS) and allowed some of the better no line of sight frequency (700 MHz) into the FCC Part-15 Rules. The Wireless Internet Service Providers Association, a national association of WISPs, petitioned the FCC and won.

Initially, WISPs were only found in rural areas not covered by cable or DSL. These early WISPs would employ a high-capacity T-carrier, such as a T1 or DS3 connection, and then broadcast the signal from a high elevation, such as at the top of a water tower. To receive this type of Internet connection, consumers mount a small dish to the roof of their home or office and point it to the transmitter. Line of sight is usually necessary for WISPs operating in the 2.4 and 5 GHz bands with 900 MHz offering better NLOS (non-line-of-sight) performance.

## Residential Wireless Internet

Providers of fixed wireless broadband services typically provide equipment to customers and install a small antenna or dish somewhere on the roof. This equipment is usually deployed as a service and maintained by the company providing that service. Fixed wireless services have become particularly popular in many rural areas where cable, DSL or other typical home internet services are not available.

## Business Wireless Internet

Many companies in the US and worldwide have started using wireless alternatives to incumbent and local providers for internet and voice service. These providers tend to offer competitive services and options in areas where there is a difficulty getting affordable Ethernet connections from terrestrial providers such as ATT, Comcast, Verizon and others. Also, companies looking for full diversity between carriers for critical uptime requirements may seek wireless alternatives to local options.

## Demand for Spectrum

To cope with increased demand for wireless broadband, increased spectrum would be needed. Studies began in 2009, and while some unused spectrum was available, it appeared broadcasters would have to give up at least some spectrum. This led to strong objections from the broadcasting community. In 2013, auctions were planned, and for now any action by broadcasters is voluntary.

## Mobile Wireless Broadband

Wireless broadband technologies include services from mobile phone service providers such as Verizon Wireless, Sprint Corporation, and AT&T Mobility, and T-Mobile which allow a more mobile version of Internet access. Consumers can purchase a PC card, laptop card, or USB equipment to connect their PC or laptop to the Internet via cell phone towers. This type of connection would be stable in almost any area that could also receive a strong cell phone connection. These connections can cost more for portable convenience as well as having speed limitations in all but urban environments.

On June 2, 2010, after months of discussion, AT&T became the first wireless Internet provider in the USA to announce plans to charge according to usage. As the only iPhone service in the United States, AT&T experienced the problem of heavy Internet use more than other providers. About 3 percent of AT&T smart phone customers account for 40 percent of the technology's use. 98 percent of the company's customers use less than 2 gigabytes, the limit under the $25 monthly plan, and 65 percent use less than 200 megabytes, the limit for the $15 plan. For each gigabyte in excess of the limit, customers would be charged $10 a month starting June 7, 2010, though existing customers would not be required to change from the $30 a month unlimited service plan. The new plan would become a requirement for those upgrading to the new iPhone technology later in the summer.

# WIRELESS INTEGRATED NETWORK SENSORS

Wireless integrated network sensors (WINS) provide distributed network and Internet access to sensors, controls, and processors embedded in equipment, facilities, and the environment. WINS combine sensor technology, signal processing, computation, and wireless networking capability in integrated systems. With advances in integrated circuit technology, sensors, radios, and processors can now be constructed at low cost and with low power consumption, enabling mass production of sophisticated compact systems that can link the physical world to networks. These systems can be local or global and will have many applications, including medicine, security, factory automation, environmental monitoring, and condition-based maintenance. Because of their compactness and low cost, WINS can be embedded and distributed at a small fraction of the cost of conventional wire-line sensor and actuator systems. Designers of systems with hundreds, or even thousands, of sensors will face many challenges.

Centralized methods of sensor networking make impractical demands on cable installations and network bandwidth. The burden on communication system components, networks, and human resources can be drastically reduced if raw data are processed at the source and the decisions conveyed. The same holds true for systems with relatively thin communications pipes between a source and the end network or systems with large numbers of devices. The physical world generates an unlimited quantity of data that can be observed, monitored, and controlled, but wireless telecommunications infrastructure are finite. Thus, even as mobile broadband services become available, processing of raw data at the source and careful control of communications access will be necessary.

The first example is an autonomous network of sensors used to monitor events in the physical world for the benefit of a remote user connected via the Web. The second scenario explores how sensor information from an automobile could be used. A general architecture for both is shown in the figure. The figure does not show in detail how services can actually be supported by Internet-connected devices, but two clusters of nodes, connected through separate gateways to the Internet, can supply some services. The nodes are assumed to be addressable either through an Internet protocol address or an attribute (e.g., location, type, etc.). Unlike pure networking elements, the nodes contain a combination of sensors and actuators. In other words, they interact with the physical world. The gateway may be a sensor node similar to other nodes in the cluster, or it may be entirely different, performing, for example, extra signal processing and communications tasks and having no sensors. In the cluster in the top left portion of figure, nodes are connected by a multihop network, with redundant pathways to the gateway. In the bottom cluster, nodes may be connected to the gateway through multihop wireless networks or through other means, such as a wired local area network (LAN). The nodes in different clusters may be all one type or they may vary within or among clusters. In a remote monitoring situation, part of the target region may have no infrastructure; thus, the multihop network must be capable of self-organization. Other parts of the region may already have assets in place that are accessible through a preexisting LAN. There is no requirement that these assets be either small or wired. The point is to design a system that makes use of all available devices to provide the desired service.

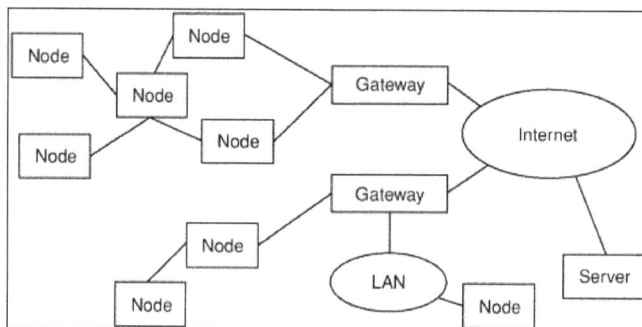

WINS network architecture.

## Design Heuristics

Pottie and Kaiser described some of the fundamental physical constraints on the cost of sensing, detection, communication, and signal processing. They identified five basic design constraints:

- For reliable detection in many situations, sensors must be in close proximity to a physical event (e.g., physical obstructions to cameras); thus large numbers of sensors may be needed. The type of information obtained with large numbers of sensors is qualitatively different from the information obtained with remote arrays.

- The cost of sensors, radios, and signal processing will come down as the cost of integrated circuit technology comes down. The cost of batteries and other energy sources, however, will come down much more slowly.

- The cost per bit for communications energy is often many orders of magnitude higher than for the energy required to make decisions at the source. Whereas processing cost is limited to first order only by current technology, the efficiency of communications has fundamental limits.

- Networks must be self-organizing to be economical.

- Scaling to larger numbers while maintaining physical responsiveness requires a hierarchy with distributed operation at lower levels and increasingly centralized control at higher levels.

Note that a hierarchy does not necessarily imply a need for heterogeneous devices. Consider, for example, a human organization. The processing abilities are roughly equal at all levels, but in progressing up the chain, different information is processed at different levels of abstraction and aggregation. Commands moving down the chain also differ in the level of abstraction, from policies to work directives that require different levels of interpretation. This flexibility enables individuals at the lower levels to deal with local changes in the work situation much faster than if a central controller had to be consulted for each action; at the same time, global goals continue to be pursued. With machines of course, we can provide highly differentiated abilities to devices at different levels of the hierarchy. For example, a backbone long-range high-speed communications pipe can greatly reduce latency compared to multihop links. Thus, even though a logical rather than physical hierarchy is arguably much more important to scalability, the designer of large-scale systems must not be seduced by the siren song of homogeneity and should consider both. In any case, homogeneity is impractical in long-lived systems composed of integrated circuit components. For systems that use the Internet, the architecture must accommodate successive generations of more powerful components.

## Remote Monitoring

The targets could be military vehicles, species of animals, pollutants, seismic events on Mars, or, on a smaller scale, enzyme levels in the bloodstream. In any case, let's assume there is no local power grid or wired communications infrastructure, but that there are long-range communications for getting information to and from a remote user. Both energy and communications bandwidth can be critical constraints. If the network must scale in the number of elements, much of the signal processing will have to be performed locally. For example, in studying the behavior of animals in the wild, a dense network of acoustic sensors might be used. The nodes would contain templates for identifying the species emitting the call. Nodes that made a tentative identification could then alert their immediate neighbors so the location of the animal could be roughly determined by triangulation. Infrared and seismic sensors might also be used in the initial identification and location processes.

Other nodes would then be activated to take a picture of the target location so a positive identification could be made. This hierarchy of signal processing and communications would be orders of magnitude more efficient in terms and energy and bandwidth than sending images of the entire region to the gateway. In addition, with the interaction of different types of nodes, most of the monitoring would be automated; humans would be brought into the loop only for the difficult final recognition of the visual pattern of prese-lected images. Upon positive identification, the audio and infrared files corresponding to the image would be added to a database, which could subsequently be mined to pro-duce better identification templates. Note that with long-range communications links (via the gateway), the user could make the full use of web-accessible utilities. Thus the end user would not have to be present in the remote location, and databases, computing resources, and the like could all be brought to bear on interpreting the (processed) data.

Experimental apparatus for initial exploration of an application domain and the apparatus that will actually be needed for large-scale deployment may differ. Because networked sensors have hitherto been very expensive, relatively little array data are available for most identification purposes, and sensors have typically been placed much farther from potential targets than they will be with WINS. This means, paradoxically, that initially fairly powerful nodes will have to be constructed to conduct large-scale experiments to collect raw data and suitable identification algorithms developed from the resulting database. In experimenting with different networking algorithms, it is desirable, from the point of view of software development, to provide an initial platform with considerable flexibility. The DARPA SensIT Program has produced development platforms to support this kind of experimentation. Other researchers have focused on specializing functions and miniaturizing components to demonstrate that large networks of small nodes can be produced. Sensoria Corporation's WINS NG 2.0, for example, nodes include ports for four sensors, a realtime digital signal processor, memory, a main processor running Linux, a battery and port for external power, the global positioning system (GPS), Ethernet, an RS-232 port, and two radios for convenient synthesis of multihop networks. Software interfaces have been created to enable programmers to control remotely a large number of physical attributes of nodes and to download new applications remotely. Thus, diverse users can produce algorithms for networking, target identification, and distributed database management. On another track, researchers at the University of California-Berkeley are engaged in producing very small nodes with limited sensing and communications abilities to demonstrate that sensing, signal processing, and communications can be combined in a miniature package.

## Automotive Applications

All automobiles produced recently include many processors and sensors, as well as a variety of networks for sensing, control, and entertainment systems. For example, hundreds of sensor parameters are accessible through the on-board diagnostic port. However, there are no connections between these networks and external communications systems, such as cellular phones. The Automotive Multimedia Interface Collaboration (AMI-C) has been working on ways to connect these networks and provide standardized buses in automobiles so a wide range of consumer electronics can be installed. This would create an automotive intranet that could then be conveniently accessed via the Internet. Ports on the bus could include any of a number of radios, so wireless devices in the vehicle could become part of the intranet, or short-range high-speed communications could be possible between a vehicle and a residence or service station.

A key component of the architecture envisioned in the AMI-C standard is a gateway that separates proprietary and safety-sensitive systems in the automobile from after-market consumer electronics. The gateway would have separate ports for interfacing with legacy networks and consumer buses. The gateway would also host software for managing the various services envisioned for internet-connected vehicles. For example, maintenance information would enable manufacturers to learn how their vehicles are actually used or enable consumers to evaluate the need for repairs and determine the effectiveness of repairs by comparing data before and after. Other potential uses could include uploading of entertainment information and locating nearby retail stores, restaurants, or service stations.

A vital function of the gateway in making such services economical is management of the communications links. Presently, cell phones have a much higher cost per bit delivered than other means of communication. However, if the automobile also has a short-range broadband link, such as IEEE 802.11b, then information might be processed and stored until it can be uploaded to a home computer when the car is parked near the residence. In a similar way, entertainment information or software upgrades could be downloaded overnight. Another approach would be to communicate over high-speed links at a gas station during refueling, for example, to receive updated information or complete a purchase of digital audio files. For very high-priority services, such as emergency assistance, the cell phone would be used, rather than waiting until a high-speed port comes into range. Based on the vehicle operator's preferences, the gateway could choose an appropriate mix of local processing, storage, and communications that would provide services at the desired costs.

The high-level requirements for the design of the gateway are surprisingly similar to the requirements for the development nodes described in the first scenario for conducting large-scale data collection experiments. Common requirements include real-time components, general purpose processors, wired and wireless network communication interfaces, application program interfaces that permit construction of software by third parties, and remote controllability via the Web. Although the devices are quite different, the same architecture applies. For a vehicle, the gateway may have some devices that respond to the physical world directly, or such devices may be accessible through local area networks in the vehicle. For the sake of economy, some of these devices and the gateway would perform local processing. Rather than sending a continuous record of engine temperature, for example, detailed reports might be stored only when temperatures cross a critical threshold or when the temperature is high and another sensor indicates possible problems. Further processing might even make a preliminary diagnosis, after which a query to an expert system located on the Web might be made. In this way, the vehicle would not have to host the complete diagnostics system.

Remote monitoring and control would also be attractive for other reasons. Vehicle owners will probably not want to program their preferences while operating the vehicle, and any sensible regulatory regime will surely discourage driver distractions. Scaling is also a concern. Providing services to millions of vehicles presents enormous challenges, both in terms of the huge volume of data that can be generated by vehicles and the quantity of entertainment information that may have to be transported to them. With a gateway and back-end web-server network that enables remote downloading of software, many different companies will be able to compete for providing information services to automobile owners.

# WIRELESS INTERNET SERVICE PROVIDER

A wireless Internet service provider (WISP) is an Internet service provider that allows users to connect to a server through a wireless connection such as Wi-Fi. WISPs provide additional services such as virtual private networking VoIP and location-based content.

In the United States, wireless networking is mainly chosen by isolated municipal ISPs and large state-wide initiatives. WISPs are more popular in rural areas, where the users may not be able to use cable and digital subscriber lines (DSL) for Internet access.

Wireless Internet service providers mesh networking or other devices built to operate over open bands between 900 MHz and 5.8 GHz. The devices may also include licensed frequencies in ultra-high frequency (UHF) bands, including multichannel multipoint distribution service (MMDS) bands.

The operating mechanism of a WISP involves pulling an expensive and large point-to-point connection to the center of the area that needs to be serviced. The process involves scanning the area for an elevated building on which wireless equipment can be mounted. The WISP may also connect to a point-to-presence (PoP) and then backhaul to the required towers, thereby eliminating the need to provide a point-to-point connection to the tower.

For consumers who wish to access a WISP connection, a small dish or antenna is placed on the roof of the consumer's house and is pointed back to the WISP's nearest antenna site. In a heavily populated area operating at 2.4 GHz band frequency, access points mounted on light posts and consumer buildings can be quite common.

It is often difficult for a single service provider to invest in building an infrastructure to offer global access to its users. In order to encourage roaming between service providers, a Wi-Fi alliance has been established, which approves a set of recommendations known as WISPr to enable internetwork and interoperator roaming for Wi-Fi users.

A wireless Internet service provider (WISP) is an Internet service provider with a network based on wireless networking. Technology may include commonplace Wi-Fi wireless mesh networking, or proprietary equipment designed to operate over open 900 MHz, 2.4 GHz, 4.9, 5, 24, and 60 GHz bands or licensed frequencies in the UHF band (including the MMDS frequency band), LMDS, and other bands from 6Ghz to 80Ghz.

In the US, the Federal Communications Commission (FCC) released Report and Order, FCC 05-56 in 2005 that revised the FCC's rules to open the 3650 MHz band for terrestrial wireless broadband operations. On November 14, 2007 the Commission released Public Notice (DA 07-4605) in which the Wireless Telecommunications Bureau announced the start date for licensing and registration process for the 3650-3700 MHz band.

As of July 2015, there are over 1,280 fixed wireless broadband providers operating in the US covering 51% of the US population.

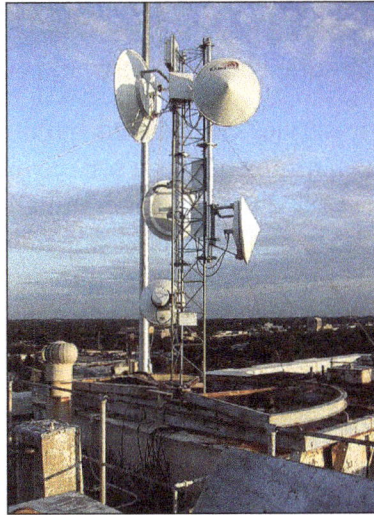

Aspen Communication's wireless access point in Tyler.

An embedded RouterBoard 112 with U.FL-RSMA pigtail and R52 miniPCI Wi-Fi card widely used by WISPs in the Czech Republic.

WISPs often offer additional services like location-based content, Virtual Private Networking and Voice over IP. Isolated municipal ISPs and larger statewide initiatives alike are tightly focused on wireless networking.

WISPs have a large market share in rural environments where cable and digital subscriber lines are not available; further, with technology available, they can meet or beat speeds of legacy cable and telephone systems. In urban environments, Gigabit Wireless links are common and provide levels of bandwidth previously only available through expensive fiber optic connections.

Typically, the way that a WISP operates is to order a fiber circuit to the center of the area they wish to serve. From there, the WISP will start building backhauls (gigabit wireless or fiber) to elevated points in the region, such as a radio towers, tall buildings, grain silos, or water towers. Those locations will have access points to provide service to individual customers or backhauls to other towers where they have more equipment. The WISP may also use gigabit wireless links to connect a PoP (Point of Presence) to several towers, reducing the need to pay for fiber circuits to the tower. For fixed wireless connections, a small dish or antenna is mounted to the roof of the customer's building and aligned to the WISP's nearest antenna site. When operating over the tightly limited range of the heavily populated 2.4 GHz band, as nearly all 802.11-based WiFi providers do, it is not uncommon to also see access points mounted on light posts and customer buildings.

Since it is difficult for a single service provider to build an infrastructure that offers global access to its subscribers, roaming between service providers is encouraged by the Wi-Fi Alliance with the protocol WISPr, a set of recommendations approved by the alliance which facilitate inter-network and inter-operator roaming of Wi-Fi users. Modern wireless services have latency comparable to other terrestrial broadband networks.

## WIRELESS LAN

A wireless LAN (WLAN) is a wireless computer network that links two or more devices using wireless communication to form a local area network (LAN) within a limited area such as a home, school, computer laboratory, campus, or office building. This gives users the ability to move around within the area and remain connected to the network. Through a gateway, a WLAN can also provide a connection to the wider Internet.

Most modern WLANs are based on IEEE 802.11 standards and are marketed under the Wi-Fi brand name.

Wireless LANs have become popular for use in the home, due to their ease of installation and use. They are also popular in commercial properties that offer wireless access to their employees and customers.

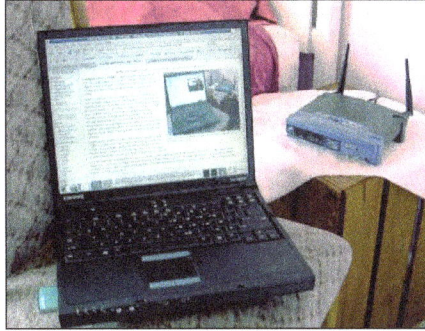

This notebook computer is connected to a wireless access point using a PC Card wireless card.

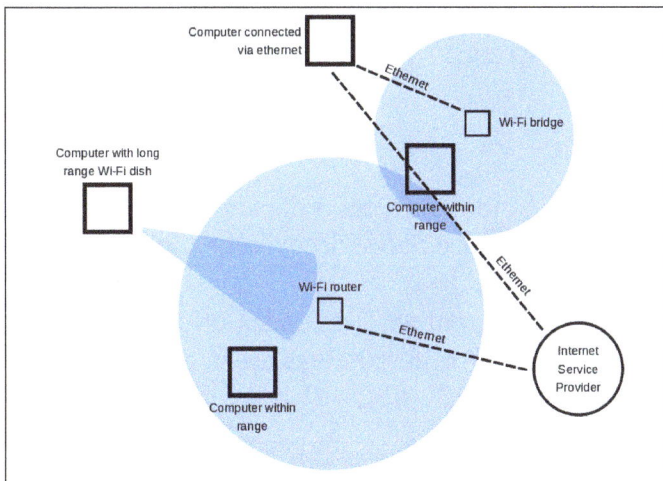

An example of a Wi-Fi network.

## Architecture

### Stations

All components that can connect into a wireless medium in a network are referred to as stations (STA). All stations are equipped with wireless network interface controllers (WNICs). Wireless stations fall into two categories: wireless access points, and clients. Access points (APs), normally wireless routers, are base stations for the wireless network. They transmit and receive radio frequencies for wireless enabled devices to communicate with. Wireless clients can be mobile devices such as laptops, personal digital assistants, IP phones and other smartphones, or non-portable devices such as desktop computers, printers, and workstations that are equipped with a wireless network interface.

### Basic Service Set

The basic service set (BSS) is a set of all stations that can communicate with each other at PHY layer. Every BSS has an identification (ID) called the BSSID, which is the MAC address of the access point servicing the BSS.

There are two types of BSS: Independent BSS (also referred to as IBSS), and infrastructure BSS. An independent BSS (IBSS) is an ad hoc network that contains no access points, which means they cannot connect to any other basic service set.

### Independent Basic Service Set

An IBSS is a set of STAs configured in ad hoc (peer-to-peer)mode.

### Extended Service Set

An extended service set (ESS) is a set of connected BSSs. Access points in an ESS are connected by a distribution system. Each ESS has an ID called the SSID which is a 32-byte (maximum) character string.

### Distribution System

A distribution system (DS) connects access points in an extended service set. The concept of a DS can be used to increase network coverage through roaming between cells.

DS can be wired or wireless. Current wireless distribution systems are mostly based on WDS or MESH protocols, though other systems are in use.

### Types of Wireless LANs

The IEEE 802.11 has two basic modes of operation: infrastructure and *ad hoc* mode. In *ad hoc* mode, mobile units transmit directly peer-to-peer. In infrastructure mode, mobile units communicate through an access point that serves as a bridge to other networks (such as Internet or LAN).

Since wireless communication uses a more open medium for communication in comparison to wired LANs, the 802.11 designers also included encryption mechanisms: Wired Equivalent Privacy (WEP, now insecure), Wi-Fi Protected Access (WPA, WPA2, WPA3), to secure wireless computer networks. Many access points will also offer Wi-Fi Protected Setup, a quick (but now insecure) method of joining a new device to an encrypted network.

### Infrastructure

Most Wi-Fi networks are deployed in infrastructure mode. In infrastructure mode, a base station acts as a wireless access point hub, and nodes communicate through the hub. The hub usually, but not always, has a wired or fiber network connection, and may have permanent wireless connections to other nodes.

Wireless access points are usually fixed, and provide service to their client nodes within range.

Wireless clients, such as laptops and smartphones, connect to the access point to join the network.

Sometimes a network will have a multiple access points, with the same 'SSID' and security arrangement. In that case connecting to any access point on that network joins the client to the network. In that case, the client software will try to choose the access point to try to give the best service, such as the access point with the strongest signal.

## Peer-to-peer

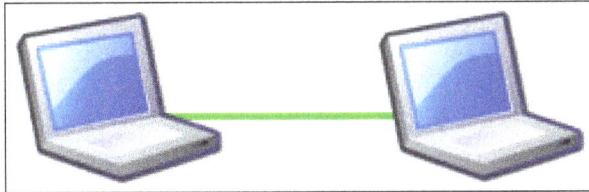

Peer-to-Peer or ad hoc wireless LAN.

An ad hoc network (not the same as a WiFi Direct network) is a network where stations communicate only peer to peer (P2P). There is no base and no one gives permission to talk. This is accomplished using the Independent Basic Service Set (IBSS).

A WiFi Direct network is another type of network where stations communicate peer to peer.

In a Wi-Fi P2P group, the group owner operates as an access point and all other devices are clients. There are two main methods to establish a group owner in the Wi-Fi Direct group. In one approach, the user sets up a P2P group owner manually. This method is also known as Autonomous Group Owner (autonomous GO). In the second method, also called negotiation-based group creation, two devices compete based on the group owner intent value. The device with higher intent value becomes a group owner and the second device becomes a client. Group owner intent value can depend on whether the wireless device performs a cross-connection between an infrastructure WLAN service and a P2P group, remaining power in the wireless device, whether the wireless device is already a group owner in another group and a received signal strength of the first wireless device.

A peer-to-peer network allows wireless devices to directly communicate with each other. Wireless devices within range of each other can discover and communicate directly without involving central access points. This method is typically used by two computers so that they can connect to each other to form a network. This can basically occur in devices within a closed range.

If a signal strength meter is used in this situation, it may not read the strength accurately and can be misleading, because it registers the strength of the strongest signal, which may be the closest computer.

IEEE 802.11 defines the physical layer (PHY) and MAC (Media Access Control) layers based on CSMA/CA (Carrier Sense Multiple Access with Collision Avoidance). This is in contrast to Ethernet which uses CSMA-CD (Carrier Sense Multiple Access with Collision Detection). The 802.11 specification includes provisions designed to minimize collisions, because two mobile units may both be in range of a common access point, but out of range of each other.

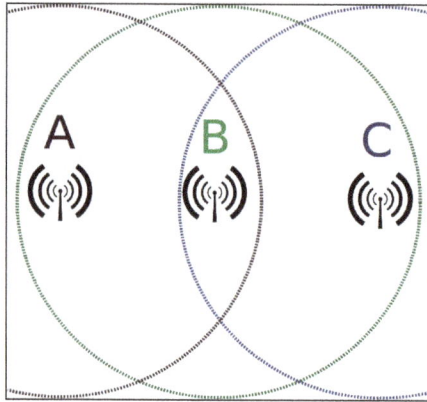

Hidden node problem: Devices A and C are both
communicating with B, but are unaware of each other.

## Bridge

A bridge can be used to connect networks, typically of different types. A wireless Ethernet bridge allows the connection of devices on a wired Ethernet network to a wireless network. The bridge acts as the connection point to the Wireless LAN.

## Wireless Distribution System

A wireless distribution system (WDS) enables the wireless interconnection of access points in an IEEE 802.11 network. It allows a wireless network to be expanded using multiple access points without the need for a wired backbone to link them, as is traditionally required. The notable advantage of a WDS over other solutions is that it preserves the MAC addresses of client packets across links between access points.

An access point can be either a main, relay or remote base station. A main base station is typically connected to the wired Ethernet. A relay base station relays data between remote base stations, wireless clients or other relay stations to either a main or another relay base station. A remote base station accepts connections from wireless clients and passes them to relay or main stations. Connections between clients are made using MAC addresses rather than by specifying IP assignments.

All base stations in a WDS must be configured to use the same radio channel, and share WEP keys or WPA keys if they are used. They can be configured to different service set identifiers. WDS also requires that every base station be configured to forward to others in the system.

WDS capability may also be referred to as repeater mode because it appears to bridge and accept wireless clients at the same time (unlike traditional bridging). Throughput in this method is halved for all clients connected wirelessly.

When it is difficult to connect all of the access points in a network by wires, it is also possible to put up access points as repeaters.

## Roaming

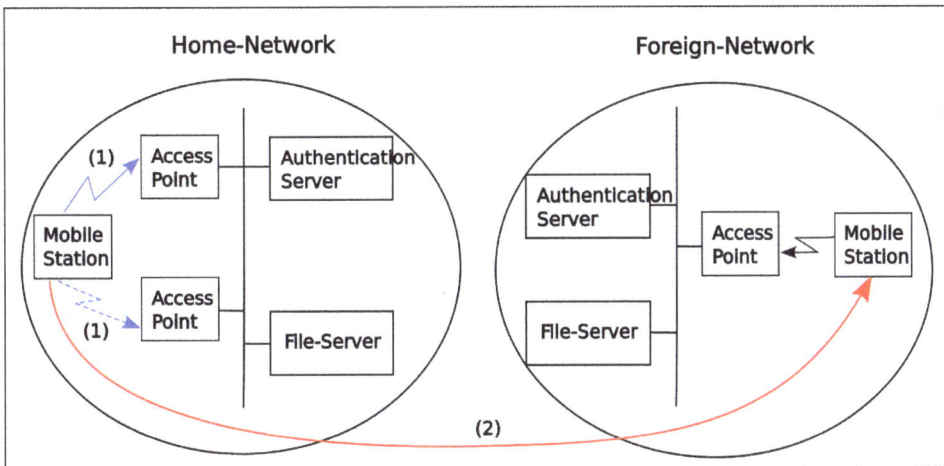

Roaming among Wireless Local Area Networks.

There are two definitions for wireless LAN roaming:

- Internal roaming: The Mobile Station (MS) moves from one access point (AP) to another AP within a home network if the signal strength is too weak. An authentication server (RADIUS) performs the re-authentication of MS via 802.1x (e.g. with PEAP). The billing of QoS is in the home network. A Mobile Station roaming from one access point to another often interrupts the flow of data among the Mobile Station and an application connected to the network. The Mobile Station, for instance, periodically monitors the presence of alternative access points (ones that will provide a better connection). At some point, based on proprietary mechanisms, the Mobile Station decides to re-associate with an access point having a stronger wireless signal. The Mobile Station, however, may lose a connection with an access point before associating with another access point. In order to provide reliable connections with applications, the Mobile Station must generally include software that provides session persistence.

- External roaming: The MS (client) moves into a WLAN of another Wireless Internet Service Provider (WISP) and takes their services (Hotspot). The user can use a foreign network independently from their home network, provided that the foreign network allows visiting users on their network. There must be special authentication and billing systems for mobile services in a foreign network.

## Applications

Wireless LANs have a great deal of applications. Modern implementations of WLANs range from small in-home networks to large, campus-sized ones to completely mobile networks on airplanes and trains.

Users can access the Internet from WLAN hotspots in restaurants, hotels, and now with portable devices that connect to 3G or 4G networks. Oftentimes these types of public access points require no registration or password to join the network. Others can be accessed once registration has occurred and a fee is paid.

Existing Wireless LAN infrastructures can also be used to work as indoor positioning systems with no modification to the existing hardware.

## Wireless LAN controller

A wireless LAN (or WLAN) controller is used in combination with the Lightweight Access Point Protocol (LWAPP) to manage light-weight access points in large quantities by the network administrator or network operations center. The wireless LAN controller is part of the Data Plane within the Cisco Wireless Model. The WLAN controller automatically handles the configuration of wireless access-points.

## Features

- Interference detection and avoidance: RF power and channel assignment will be adjusted to the plan.

- Load balancing: Disabled by default, high-speed load balancing can be used to connect an user to multiple access points for better coverage and data rates.

- Coverage hole detection and correction: Part of the RF management is the ability to handle power levels. Power can be increased to cover holes or reduced to protect against cell overlapping.

The WLAN controller also comes with ' various forms of authentication such as: 802.1X (Protected Extensible Authentication Protocol (PEAP), LEAP, EAP-TLS, Wi-Fi Protected Access (WPA), 802.11i (WPA2), and Layer 2 Tunneling Protocol (L2TP).

# MOBILITY MANAGEMENT

Mobility management is a functionality that facilitates mobile device operations in Universal Mobile Telecommunications System (UMTS) or Global System for Mobile Communications (GSM) networks. Mobility management is used to trace physical user and subscriber locations to provide mobile phone services, like calls and Short Message Service (SMS).

UMTS and GSM are each made up of separate cells (base stations) that cover a specific geographical area. All base stations are integrated into one area, allowing a cellular network to cover a wider area (location area).

The location update procedure allows a mobile device to notify a cellular network when shifting between areas. When a mobile device recognizes that an area code differs from a previous update, the mobile device executes a location update, by sending a location request to its network, prior location and specific Temporary Mobile Subscriber Identity (TMSI). A mobile device provides updated network location information for several reasons, including reselecting cell location coverage due to a faded signal.

Location area includes a group of base stations assembled collectively to optimize signaling. Base stations are integrated to form a single network area known as a base station controller (BSC). The BSC manages allocation of radio channels, acquires measurements from cell phones, and handles handovers from one base station to another.

Roaming is among the basic procedures of mobility management. It enables subscribers to use mobile services when moving outside of the geographical area of a specific network.

## Location Update Procedure

A GSM or UMTS network, like all cellular networks, is basically a radio network of individual cells, known as base stations. Each base station covers a small geographical area which is part of a uniquely identified location area. By integrating the coverage of each of these base stations, a cellular network provides a radio coverage over a much wider area. For GSM, a base station is called a base transceiver station (BTS), and for UMTS it is called a Node B. A group of base stations is named a location area, or a routing area.

The location update procedure allows a mobile device to inform the cellular network, whenever it moves from one location area to the next. Mobiles are responsible for detecting location area codes (LAC). When a mobile finds that the location area code is different from its last update, it performs another update by sending to the network, a location update request, together with its previous location, and its Temporary Mobile Subscriber Identity (TMSI).

The mobile also stores the current LAC in the SIM card, concatenating it to a list of recently used LACs. This is done to avoid unnecessary IMSI attachment procedures in case the mobile has been forced to switch off (by removing the battery, for example) without having a chance to notify the network with an IMSI detach and then switched on right after it has been turned off. Considering the fact that the mobile is still associated with the Mobile Switching Center/Visitor Location Register (MSC/VLR) of the current location area, there is no need for any kind of IMSI attachment procedures to be done.

There are several reasons why a mobile may provide updated location information to the network. Whenever a mobile is switched on or off, the network may require it to perform an IMSI attach or IMSI detach location update procedure. Also, each mobile is required to regularly report its location at a set time interval using a *periodic location update* procedure. Whenever a mobile moves from one location area to the next while not on a call, a *random location* update is required. This is also required of a stationary mobile that reselects coverage from a cell in a different location area, because of signal fade. Thus, a subscriber has reliable access to the network and may be reached with a call, while enjoying the freedom of mobility within the whole coverage area.

When a subscriber is paged in an attempt to deliver a call or SMS and the subscriber does not reply to that page then the subscriber is marked as absent in both the MSC/VLR and the Home Location Register (HLR) (Mobile not reachable flag MNRF is set). The next time the mobile performs a location update, the HLR is updated and the mobile not reachable flag is cleared.

## TMSI

The Temporary Mobile Subscriber Identity (TMSI) is the identity that is most commonly sent between the mobile and the network. TMSI is randomly assigned by the VLR to every mobile in the area, the moment it is switched on. The number is local to a location area, and so it has to be updated each time the mobile moves to a new geographical area.

The network can also change the TMSI of the mobile at any time. And it normally does so, in order to avoid the subscriber from being identified, and tracked by eavesdroppers on the radio interface. This makes it difficult to trace which mobile is which, except briefly, when the mobile is just switched on, or when the data in the mobile becomes invalid for one reason or another. At that point, the global "international mobile subscriber identity" (IMSI) must be sent to the network. The IMSI is sent as rarely as possible, to avoid it being identified and tracked.

A key use of the TMSI is in paging a mobile. "Paging" is the one-to-one communication between the mobile and the base station. The most important use of broadcast information is to set up channels for "paging". Every cellular system has a broadcast mechanism to distribute such information to a plurality of mobiles.

Size of TMSI is 4 octet with full hex digits and can't be all FF because the SIM uses 4 octets with all bits equal to 1 to indicate that no valid TMSI is available.

## Roaming

Roaming is one of the fundamental mobility management procedures of all cellular networks. Roaming is defined as the ability for a cellular customer to automatically make and receive voice calls, send and receive data, or access other services, including home data services, when travelling outside the geographical coverage area of the home network, by means of using a visited network. This can be done by using a communication terminal or else just by using the subscriber identity in the visited network. Roaming is technically supported by a mobility management, authentication, authorization and billing procedures.

## Types of Area

### Location Area

A "location area" is a set of base stations that are grouped together to optimize signaling. Typically, tens or even hundreds of base stations share a single Base Station Controller (BSC) in GSM, or a Radio Network Controller (RNC) in UMTS. The BSC/RNC is the intelligence behind the base stations; it handles allocation of radio channels, receives measurements from the mobile phones, and controls handovers from base station to base station.

To each location area, a unique number called a "location area code" (LAC) is assigned. The location area code is broadcast by each base station at regular intervals. Within a location area, each base station is assigned a distinct "cell identifier" (CI) number.

If the location areas are very large, there will be many mobiles operating simultaneously, resulting in very high paging traffic, as every paging request has to be broadcast to every base station in the location area. This wastes bandwidth and power on the mobile, by requiring it to listen for broadcast messages too much of the time. If on the other hand, there are too many small location areas, the mobile must contact the network very often for changes of location, which will also drain the mobile's battery. A balance has therefore to be struck.

### Routing Area

The routing area is the packet-switched domain equivalent of the location area. A "routing area" is normally a subdivision of a "location area". Routing areas are used by mobiles which are GPRS-attached. GPRS is optimized for "bursty" data communication services, such as wireless internet/intranet, and multimedia services. It is also known as GSM-IP ("Internet Protocol") because it will connect users directly to Internet Service Providers.

The bursty nature of packet traffic means that more paging messages are expected per mobile, and so it is worth knowing the location of the mobile more accurately than it would be with traditional circuit-switched traffic. A change from routing area to routing area (called a "Routing Area Update") is done in an almost identical way to a change from location area to location area. The main differences are that the "Serving GPRS Support Node" (SGSN) is the element involved.

## Tracking Area

The tracking area is the LTE counterpart of the location area and routing area. A tracking area is a set of cells. Tracking areas can be grouped into lists of tracking areas (TA lists), which can be configured on the User Equipment (UE). Tracking area updates are performed periodically or when the UE moves to a tracking area that is not included in its TA list.

Operators can allocate different TA lists to different UEs. This can avoid signaling peaks in some conditions: for instance, the UEs of passengers of a train may not perform tracking area updates simultaneously.

On the network side, the involved element is the Mobility Management Entity (MME). MME configures TA lists using NAS messages like Attach Accept, TAU Accept or GUTI Reallocation Command.

## Wireless Mobility Management

Wireless mobility management in Personal Communications Service (PCS) is the assigning and controlling of wireless links for terminal network connections. Wireless mobility management provides an "alerting" function for call completion to a wireless terminal, monitors wireless link performance to determine when an automatic link transfer is required, and coordinates link transfers between wireless access interfaces.

One use of this is wireless push technology, by pushing data across wireless networks, this coordinates the link transfers and pushes data between the backend and wireless device only when an established connection is found.

# WIRELESS NETWORK AFTER NEXT

The Wireless Network after Next was a DARPA project to create and demonstrate an advanced tactical mobile ad-hoc network that rapidly adapts to soldiers maneuvering in complicated environments, automatically determining the best radio frequencies and network path to maximize connectivity and throughput. In 2010 it was successfully demonstrated in live military experiments containing up to 100 nodes, the largest military MANET demonstrated to that date.

The Wireless Network after Next consisted of a novel radio platform, created by Cobham plc, and novel networking protocols, created by BBN Technologies:

- The WNaN radio was envisioned as a low-cost, hand-held, multi-channel, spectrum-agile, MIMO-capable wireless node, built with inexpensive RF circuit technology. It operated in the 900 MHz to 6 GHz frequency band, and contained multiple radio transceivers for simultaneous transmission and reception across multiple frequency bands at rates greater than 1 megabit/second.

- The WNaN network provided highly adaptive communications in a self-organizing, self-healing network. Each radio node acted as a router to automatically perform dynamic spectrum management for spectrum agility, Disruption Tolerant Networking for resilience, and content-based networking for efficient data dissemination. The routers also took full advantage of sending and receiving communications packets in parallel across multiple radio transceivers.

# WIRELESS INTELLIGENT NETWORK

Wireless Intelligent Network (WIN) is a concept being developed by the Telecommunications Industry Association (TIA) Standards Committee TR45.2. The charter of this committee is to drive intelligent network (IN) capabilities, based on interim standard (IS)-41, into wireless networks. IS-41 is a standard currently being embraced by wireless providers because it facilitates roaming. Basing WIN standards on this protocol enables a graceful evolution to an IN without making current network infrastructure obsolete.

Today's wireless subscribers are much more sophisticated telecommunications users than they were five years ago. No longer satisfied with just completing a clear call, today's subscribers demand innovative ways to use the wireless phone. They want multiple services that allow them to handle or select incoming calls in a variety of ways.

Enhanced services are very important to wireless customers. They have come to expect, for instance, services such as caller ID and voice messaging bundled in the package when they buy and activate a cellular or personal communications service (PCS) phone. Whether prepaid, voice/data messaging, Internet surfing, or location-sensitive billing, enhanced services will become an important differentiator in an already crowded, competitive service-provider market.

Enhanced services will also entice potentially new subscribers to sign up for service and will drive up airtime through increased usage of PCS or cellular services. As the wireless market becomes increasingly competitive, rapid deployment of enhanced services becomes critical to a successful wireless strategy.

Intelligent network (IN) solutions have revolutionized wireline networks. Rapid creation and deployment of services has become the hallmark of a wireline network based on IN concepts. Wireless intelligent network (WIN) will bring those same successful strategies into the wireless networks.

## Benefits of Intelligent Networks

The main benefit of intelligent networks is the ability to improve existing services and develop new sources of revenue. To meet these objectives, providers require the ability to accomplish the following:

- Introduce new services rapidly: IN provides the capability to provision new services or modify existing services throughout the network with physical intervention.

- Provide service customization: Service providers require the ability to change the service logic rapidly and efficiently. Customers are also demanding control of their own services to meet their individual needs.

- Establish vendor independence: A major criterion for service providers is that the software must be developed quickly and inexpensively. To accomplish this, suppliers must integrate commercially available software to create the applications required by service providers.

- Create open interfaces: Open interfaces allow service providers to introduce network elements quickly for individualized customer services. AIN technology uses the embedded base of stored program-controlled switching systems and the SS7 network. The AIN technology also allows for the separation of service-specific functions and data from other network resources. This feature reduces the dependency on switching system vendors for software development and delivery schedules. Service providers have more freedom to create and customize services.

The SCP contains programmable service-independent capabilities (or service logic) that is under the control of service providers. The SCP also contains service-specific data that allows service providers and their customers to customize services. AIN is a logical technology, which can help service providers meet local number portability. AIN LNP solutions are so flexible that AIN provides service without the burden of costly network additions to the service providers.

## WIN Services

WIN services are related to AIN services. AIN was first introduced for the wireline industry in late 1980's. The best known AIN application is the "800 service" which opened the door to a host of new services offered on a platform other than the switch. WIN, enhancing the AIN concept with the mobility management aspect of wireless communication, will offer services consistent with what wireline AIN offers. Furthermore, WIN still needs to address:

- Personal and terminal mobility.

- Internetwork handoff.

- Security.

- Fraud prevention and detection.

## Hands-free and Voice-controlled Services

Voice-controlled services employ voice-recognition technology to allow the wireless user to control features and services using spoken commands, names, and numbers. There are two main types of automatic speech recognition (ASR). Speaker-dependent requires specific spoken phrases unique to an individual user. Each user is required to train the ASR system by recording samples of each specific phrase. The other is speaker-independent ASR, which requires the use of specific spoken phrases that are independent of the speaker. The individual user need not train the system.

## Voice Controlled Dialing (VCD)

VCD allows a subscriber to originate calls by dialing digits using spoken commands instead of the keypad. VCD may be used during call origination or during the call itself.

## Voice-controlled Feature Control (VCFC)

VCFC permits a calling party to call a special VCFC directory number, identify the calling party as an authorized subscriber with a mobile directory number and personal identification number (PIN), and specify feature operations via one or more feature-control strings. This service is similar to remote feature control (RFC) except that the subscriber is allowed to dial feature-control digits or commands using spoken words and phrases instead of keypad digits.

## Voice-based User Identification (VUI)

VUI permits a subscriber to place restrictions on access to services by using VUI to validate the identity of the speaker. VUI employs a form of ASR technology to validate the identity of the speaker rather than determine what was said by the speaker. VUI requires that the subscriber register the service by training the ASR system by recording a word or phrase. When a user attempts to access a service, the ASR system prompts the user to say the special phrase.

## Incoming Call-restriction/Control

Incoming calls to a subscriber may be given one of the following termination treatments: the call is terminated normally to the subscriber with normal or distinctive alerting; it is forwarded to voice mail or to another number; it is routed to a subscriber-specific announcement; or it is blocked. These kinds of services help subscribers control incoming calls and their monthly airtime bills. From a marketing standpoint, they entice cost-conscious customers who might not want unlimited access from callers.

# BLUETOOTH

Bluetooth is a wireless technology standard used for exchanging data between fixed and mobile devices over short distances using short-wavelength UHF radio waves in the industrial, scientific and medical radio bands, from 2.400 to 2.485 GHz, and building personal area networks (PANs). It was originally conceived as a wireless alternative to RS-232 data cables.

Bluetooth is managed by the Bluetooth Special Interest Group (SIG), which has more than 35,000 member companies in the areas of telecommunication, computing, networking, and consumer electronics. The IEEE standardized Bluetooth as IEEE 802.15.1, but no longer maintains the standard. The Bluetooth SIG oversees development of the specification, manages the qualification program, and protects the trademarks. A manufacturer must meet Bluetooth SIG standards to market it as a Bluetooth device. A network of patents apply to the technology, which are licensed to individual qualifying devices. As of 2009, Bluetooth integrated circuit chips ship approximately 920 million units annually.

## Implementation

Bluetooth operates at frequencies between 2.402 and 2.480 GHz, or 2.400 and 2.4835 GHz including guard bands 2 MHz wide at the bottom end and 3.5 MHz wide at the top. This is in the globally unlicensed (but not unregulated) industrial, scientific and medical (ISM) 2.4 GHz short-range radio frequency band. Bluetooth uses a radio technology called frequency-hopping spread spectrum. Bluetooth divides transmitted data into packets, and transmits each packet on one of 79 designated Bluetooth channels. Each channel has a bandwidth of 1 MHz. It usually performs 1600 hops per second, with adaptive frequency-hopping (AFH) enabled. Bluetooth Low Energy uses 2 MHz spacing, which accommodates 40 channels.

Originally, Gaussian frequency-shift keying (GFSK) modulation was the only modulation scheme available. Since the introduction of Bluetooth 2.0+EDR, $\pi/4$-DQPSK (differential quadrature phase-shift keying) and 8-DPSK modulation may also be used between compatible devices. Devices functioning with GFSK are said to be operating in basic rate (BR) mode where an instantaneous bit rate of 1 Mbit/s is possible. The term Enhanced Data Rate (EDR) is used to describe $\pi/4$-DPSK and 8-DPSK schemes, each giving 2 and 3 Mbit/s respectively. The combination of these (BR and EDR) modes in Bluetooth radio technology is classified as a BR/EDR radio.

Bluetooth is a packet-based protocol with a master/slave architecture. One master may communicate with up to seven slaves in a piconet. All devices share the master's clock. Packet exchange is based on the basic clock, defined by the master, which ticks at 312.5 µs intervals. Two clock ticks make up a slot of 625 µs, and two slots make up a slot pair of 1250 µs. In the simple case of single-slot packets, the master transmits in even slots and receives in odd slots. The slave, conversely, receives in even slots and transmits in odd slots. Packets may be 1, 3 or 5 slots long, but in all cases the master's transmission begins in even slots and the slave's in odd slots.

The above excludes Bluetooth Low Energy, introduced in the 4.0 specification, which uses the same spectrum but somewhat differently.

## Communication and Connection

A master BR/EDR Bluetooth device can communicate with a maximum of seven devices in a piconet (an ad-hoc computer network using Bluetooth technology), though not all devices reach this maximum. The devices can switch roles, by agreement, and the slave can become the master (for example, a headset initiating a connection to a phone necessarily begins as master—as an initiator of the connection—but may subsequently operate as the slave).

The Bluetooth Core Specification provides for the connection of two or more piconets to form a scatternet, in which certain devices simultaneously play the master role in one piconet and the slave role in another.

At any given time, data can be transferred between the master and one other device (except for the little-used broadcast mode). The master chooses which slave device to address; typically, it switches rapidly from one device to another in a round-robin fashion. Since it is the master that chooses which slave to address, whereas a slave is (in theory) supposed to listen in each receive slot, being a master is a lighter burden than being a slave. Being a master of seven slaves is possible; being a slave of more than one master is possible. The specification is vague as to required behavior in scatternets.

## Uses

| Ranges of Bluetooth devices by class | | | |
|---|---|---|---|
| Class | Max. permitted power | | Typ. range (m) |
| | (mW) | (dBm) | |
| 1 | 100 | 20 | ~100 |
| 1.5 | 10 | 10 | ~20 |
| 2 | 2.5 | 4 | ~10 |
| 3 | 1 | 0 | ~1 |
| 4 | 0.5 | −3 | ~0.5 |

Bluetooth is a standard wire-replacement communications protocol primarily designed for low power consumption, with a short range based on low-cost transceiver microchips in each device. Because the devices use a radio (broadcast) communications system, they do not have to be in visual line of sight of each other; however, a quasi optical wireless path must be viable. Range is power-class-dependent, but effective ranges vary in practice.

Officially Class 3 radios have a range of up to 1 metre (3 ft), Class 2, most commonly found in mobile devices, 10 metres (33 ft), and Class 1, primarily for industrial use cases,100 metres (300 ft). Bluetooth Marketing qualifies that Class 1 range is in most cases 20–30 metres (66–98 ft), and Class 2 range 5–10 metres (16–33 ft). The actual range achieved by a given link will depend on the qualities of the devices at both ends of the link, as well as the air conditions in between, and other factors.

The effective range varies depending on propagation conditions, material coverage, production sample variations, antenna configurations and battery conditions. Most Bluetooth applications are for indoor conditions, where attenuation of walls and signal fading due to signal reflections make the range far lower than specified line-of-sight ranges of the Bluetooth products.

Most Bluetooth applications are battery-powered Class 2 devices, with little difference in range whether the other end of the link is a Class 1 or Class 2 device as the lower-powered device tends to set the range limit. In some cases the effective range of the data link can be extended when a Class 2 device is connecting to a Class 1 transceiver with both higher sensitivity and transmission power than a typical Class 2 device. Mostly, however, the Class 1 devices have a similar sensitivity to Class 2 devices. Connecting two Class 1 devices with both high sensitivity and high power can allow ranges far in excess of the typical 100m, depending on the throughput required by the application. Some such devices allow open field ranges of up to 1 km and beyond between two similar devices without exceeding legal emission limits.

The Bluetooth Core Specification mandates a range of not less than 10 metres (33 ft), but there is no upper limit on actual range. Manufacturers' implementations can be tuned to provide the range needed for each case.

## Bluetooth Profile

To use Bluetooth wireless technology, a device must be able to interpret certain Bluetooth profiles, which are definitions of possible applications and specify general behaviors that Bluetooth-enabled devices use to communicate with other Bluetooth devices. These profiles include settings to parameterize and to control the communication from the start. Adherence to profiles saves the time for transmitting the parameters anew before the bi-directional link becomes effective. There are a wide range of Bluetooth profiles that describe many different types of applications or use cases for devices.

## List of Applications

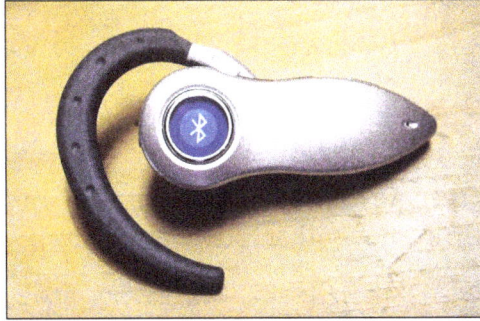

A typical Bluetooth mobile phone headset.

- Wireless control and communication between a mobile phone and a handsfree headset. This was one of the earliest applications to become popular.

- Wireless control of and communication between a mobile phone and a Bluetooth compatible car stereo system.

- Wireless communication between a smartphone and a smart lock for unlocking doors.

- Wireless control of and communication with iOS and Android device phones, tablets and portable wireless speakers.

- Wireless Bluetooth headset and Intercom. Idiomatically, a headset is sometimes called "a Bluetooth".

- Wireless streaming of audio to headphones with or without communication capabilities.

- Wireless streaming of data collected by Bluetooth-enabled fitness devices to phone or PC.

- Wireless networking between PCs in a confined space and where little bandwidth is required.

- Wireless communication with PC input and output devices, the most common being the mouse, keyboard and printer.

- Transfer of files, contact details, calendar appointments, and reminders between devices with OBEX.

- Replacement of previous wired RS-232 serial communications in test equipment, GPS receivers, medical equipment, bar code scanners, and traffic control devices.

- For controls where infrared was often used.

- For low bandwidth applications where higher USB bandwidth is not required and cable-free connection desired.

- Sending small advertisements from Bluetooth-enabled advertising hoardings to other, discoverable, Bluetooth devices.

- Wireless bridge between two Industrial Ethernet (e.g., PROFINET) networks.

- Seventh and eighth generation game consoles such as Nintendo's Wii, and Sony's PlayStation 3 use Bluetooth for their respective wireless controllers.

- Dial-up internet access on personal computers or PDAs using a data-capable mobile phone as a wireless modem.

- Short-range transmission of health sensor data from medical devices to mobile phone, set-top box or dedicated telehealth devices.

- Allowing a DECT phone to ring and answer calls on behalf of a nearby mobile phone.

- Real-time location systems (RTLS) are used to track and identify the location of objects in real time using "Nodes" or "tags" attached to, or embedded in, the objects tracked, and "Readers" that receive and process the wireless signals from these tags to determine their locations.

- Personal security application on mobile phones for prevention of theft or loss of items. The protected item has a Bluetooth marker (e.g., a tag) that is in constant communication with the phone. If the connection is broken (the marker is out of range of the phone) then an alarm is raised. This can also be used as a man overboard alarm. A product using this technology has been available since 2009.

- Calgary, Alberta, Canada's Roads Traffic division uses data collected from travelers' Bluetooth devices to predict travel times and road congestion for motorists.

- Wireless transmission of audio (a more reliable alternative to FM transmitters).

- Live video streaming to the visual cortical implant device by Nabeel Fattah in Newcastle university 2017.

- Connection of motion controllers to a PC when using VR headsets.

## Bluetooth vs Wi-Fi (IEEE 802.11)

Bluetooth and Wi-Fi (Wi-Fi is the brand name for products using IEEE 802.11 standards) have some similar applications: setting up networks, printing, or transferring files. Wi-Fi is intended as a replacement for high-speed cabling for general local area network access in work areas or home. This category of applications is sometimes called wireless local area networks (WLAN). Bluetooth was intended for portable equipment and its applications. The category of applications is outlined as the wireless personal area network (WPAN). Bluetooth is a replacement for cabling in a variety of personally carried applications in any setting, and also works for fixed location applications such as smart energy functionality in the home (thermostats, etc.).

Wi-Fi and Bluetooth are to some extent complementary in their applications and usage. Wi-Fi is usually access point-centered, with an asymmetrical client-server connection with all traffic routed through the access point, while Bluetooth is usually symmetrical, between two Bluetooth devices. Bluetooth serves well in simple applications where two devices need to connect with a minimal configuration like a button press, as in headsets and remote controls, while Wi-Fi suits better in applications where some degree of client configuration is possible and high speeds are required, especially for network access through an access node. However, Bluetooth access points do exist, and ad-hoc connections are possible with Wi-Fi though not as simply as with Bluetooth. Wi-Fi Direct was recently developed to add a more Bluetooth-like ad-hoc functionality to Wi-Fi.

## Devices

A Bluetooth USB dongle with a 100 m range.

Bluetooth exists in numerous products such as telephones, speakers, tablets, media players, robotics systems, laptops, and console gaming equipment as well as some high definition headsets, modems, hearing aids and even watches. Given the variety of devices which use the Bluetooth, coupled with the contemporary deprecation of headphone jacks by Apple, Google, and other companies, and the lack of regulation by the FCC, the technology is prone to interference. Nonetheless Bluetooth is useful when transferring information between two or more devices that are near each other in low-bandwidth situations. Bluetooth is commonly used to transfer sound data with telephones (i.e., with a Bluetooth headset) or byte data with hand-held computers (transferring files).

Bluetooth protocols simplify the discovery and setup of services between devices. Bluetooth devices can advertise all of the services they provide. This makes using services easier, because more of the security, network address and permission configuration can be automated than with many other network types.

## Computer Requirements

A typical Bluetooth USB dongle.

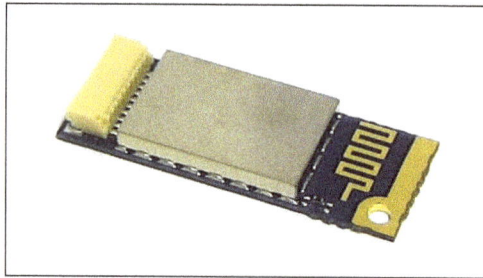

An internal notebook Bluetooth card (14×36×4 mm).

A personal computer that does not have embedded Bluetooth can use a Bluetooth adapter that enables the PC to communicate with Bluetooth devices. While some desktop computers and most recent laptops come with a built-in Bluetooth radio, others require an external adapter, typically in the form of a small USB "dongle."

Unlike its predecessor, IrDA, which requires a separate adapter for each device, Bluetooth lets multiple devices communicate with a computer over a single adapter.

## Operating System Implementation

For Microsoft platforms, Windows XP Service Pack 2 and SP3 releases work natively with Bluetooth v1.1, v2.0 and v2.0+EDR. Previous versions required users to install their Bluetooth adapter's own drivers, which were not directly supported by Microsoft. Microsoft's own Bluetooth dongles (packaged with their Bluetooth computer devices) have no external drivers and thus require at least Windows XP Service Pack 2. Windows Vista RTM/SP1 with the Feature Pack for Wireless or Windows Vista SP2 work with Bluetooth v2.1+EDR. Windows 7 works with Bluetooth v2.1+EDR and Extended Inquiry Response (EIR). The Windows XP and Windows Vista/Windows 7 Bluetooth stacks support the following Bluetooth profiles natively: PAN, SPP, DUN, HID, HCRP. The Windows XP stack can be replaced by a third party stack that supports more profiles or newer Bluetooth versions. The Windows Vista/Windows 7 Bluetooth stack supports vendor-supplied additional profiles without requiring that the Microsoft stack be replaced. It is generally recommended to install the latest vendor driver and its associated stack to be able to use the Bluetooth device at its fullest extent.

Apple products have worked with Bluetooth since Mac OS X v10.2, which was released in 2002.

Linux has two popular Bluetooth stacks, BlueZ and Fluoride. The BlueZ stack is included with most Linux kernels and was originally developed by Qualcomm. Fluoride, earlier known as Bluedroid is included in Android OS and was originally developed by Broadcom. There is also Affix stack, developed by Nokia. It was once popular, but has not been updated since 2005.

FreeBSD has included Bluetooth since its v5.0 release, implemented through netgraph.

NetBSD has included Bluetooth since its v4.0 release. Its Bluetooth stack was ported to OpenBSD as well, however OpenBSD later removed it as unmaintained.

DragonFly BSD has had NetBSD's Bluetooth implementation since 1.11. A netgraph-based implementation from FreeBSD has also been available in the tree, possibly disabled until 2014-11-15, and may require more work.

## Specifications and Features

The specifications were formalized by the Bluetooth Special Interest Group (SIG) and formally announced on the 20 of May 1998. Today it has a membership of over 30,000 companies worldwide. It was established by Ericsson, IBM, Intel, Nokia and Toshiba, and later joined by many other companies.

All versions of the Bluetooth standards support downward compatibility. That lets the latest standard cover all older versions.

The Bluetooth Core Specification Working Group (CSWG) produces mainly 4 kinds of specifications:

- The Bluetooth Core Specification, release cycle is typically a few years in between.

- Core Specification Addendum (CSA), release cycle can be as tight as a few times per year.

- Core Specification Supplements (CSS), can be released very quickly.

- Errata.

## Bluetooth 1.0 and 1.0B

Versions 1.0 and 1.0B had many problems, and manufacturers had difficulty making their products interoperable. Versions 1.0 and 1.0B also included mandatory Bluetooth hardware device address (BD_ADDR) transmission in the Connecting process (rendering anonymity impossible at the protocol level), which was a major setback for certain services planned for use in Bluetooth environments.

## Bluetooth 1.1

- Ratified as IEEE Standard 802.15.1–2002.

- Many errors found in the v1.0B specifications were fixed.

- Added possibility of non-encrypted channels.

- Received Signal Strength Indicator (RSSI).

## Bluetooth 1.2

Major enhancements include:

- Faster Connection and Discovery.

- Adaptive frequency-hopping spread spectrum (AFH), which improves resistance to radio frequency interference by avoiding the use of crowded frequencies in the hopping sequence.

- Higher transmission speeds in practice than in v1.1, up to 721 kbit/s.

- Extended Synchronous Connections (eSCO), which improve voice quality of audio links by allowing retransmissions of corrupted packets, and may optionally increase audio latency to provide better concurrent data transfer.

- Host Controller Interface (HCI) operation with three-wire UART.

- Ratified as IEEE Standard 802.15.1–2005.

- Introduced Flow Control and Retransmission Modes for L2CAP.

## Bluetooth 2.0 + EDR

This version of the Bluetooth Core Specification was released before 2005. The main difference is the introduction of an Enhanced Data Rate (EDR) for faster data transfer. The bit rate of EDR is 3 Mbit/s, although the maximum data transfer rate (allowing for inter-packet time and acknowledgements) is 2.1 Mbit/s. EDR uses a combination of GFSK and phase-shift keying modulation (PSK) with two variants, $\pi/4$-DQPSK and 8-DPSK. EDR can provide a lower power consumption through a reduced duty cycle.

The specification is published as *Bluetooth v2.0 + EDR*, which implies that EDR is an optional feature. Aside from EDR, the v2.0 specification contains other minor improvements, and products may claim compliance to "Bluetooth v2.0" without supporting the higher data rate. At least one commercial device states "Bluetooth v2.0 without EDR" on its data sheet.

## Bluetooth 2.1 + EDR

Bluetooth Core Specification Version 2.1 + EDR was adopted by the Bluetooth SIG on 26 July 2007.

The headline feature of v2.1 is secure simple pairing (SSP): this improves the pairing experience for Bluetooth devices, while increasing the use and strength of security.

Version 2.1 allows various other improvements, including extended inquiry response (EIR), which provides more information during the inquiry procedure to allow better filtering of devices before connection; and sniff subrating, which reduces the power consumption in low-power mode.

## Bluetooth 3.0 + HS

Version 3.0 + HS of the Bluetooth Core Specification was adopted by the Bluetooth SIG on 21 April 2009. Bluetooth v3.0 + HS provides theoretical data transfer speeds of up to 24 Mbit/s, though not over the Bluetooth link itself. Instead, the Bluetooth link is used for negotiation and establishment, and the high data rate traffic is carried over a colocated 802.11 link.

The main new feature is AMP (Alternative MAC/PHY), the addition of 802.11 as a high-speed transport. The high-speed part of the specification is not mandatory, and hence only devices that display the "+HS" logo actually support Bluetooth over 802.11 high-speed data transfer. A Bluetooth v3.0 device without the "+HS" suffix is only required to support features introduced in Core Specification Version 3.0 or earlier Core Specification Addendum 1.

## L2CAP Enhanced Modes

Enhanced Retransmission Mode (ERTM) implements reliable L2CAP channel, while Streaming Mode (SM) implements unreliable channel with no retransmission or flow control. Introduced in Core Specification Addendum 1.

## Alternative MAC/PHY

Enables the use of alternative MAC and PHYs for transporting Bluetooth profile data. The Bluetooth radio is still used for device discovery, initial connection and profile configuration. However, when large quantities of data must be sent, the high-speed alternative MAC PHY 802.11 (typically associated with Wi-Fi) transports the data. This means that Bluetooth uses proven low power connection models when the system is idle, and the faster radio when it must send large quantities of data. AMP links require enhanced L2CAP modes.

## Unicast Connectionless Data

Permits sending service data without establishing an explicit L2CAP channel. It is intended for use by applications that require low latency between user action and reconnection/transmission of data. This is only appropriate for small amounts of data.

## Enhanced Power Control

Updates the power control feature to remove the open loop power control, and also to clarify ambiguities in power control introduced by the new modulation schemes added for EDR. Enhanced power control removes the ambiguities by specifying the behaviour that is expected. The feature also adds closed loop power control, meaning RSSI filtering can start as the response is received. Additionally, a "go straight to maximum power" request has been introduced. This is expected to deal with the headset link loss issue typically observed when a user puts their phone into a pocket on the opposite side to the headset.

## Ultra-wideband

The high-speed (AMP) feature of Bluetooth v3.0 was originally intended for UWB, but the WiMedia Alliance, the body responsible for the flavor of UWB intended for Bluetooth, announced in March 2009 that it was disbanding, and ultimately UWB was omitted from the Core v3.0 specification.

On 16 March 2009, the WiMedia Alliance announced it was entering into technology transfer agreements for the WiMedia Ultra-wideband (UWB) specifications. WiMedia has transferred all current and future specifications, including work on future high-speed and power-optimized implementations, to the Bluetooth Special Interest Group (SIG), Wireless USB Promoter Group and the USB Implementers Forum. After successful completion of the technology transfer, marketing, and related administrative items, the WiMedia Alliance ceased operations.

In October 2009 the Bluetooth Special Interest Group suspended development of UWB as part of the alternative MAC/PHY, Bluetooth v3.0 + HS solution. A small, but significant, number of former WiMedia members had not and would not sign up to the necessary agreements for the IP transfer. The Bluetooth SIG is now in the process of evaluating other options for its longer term roadmap.

## Bluetooth 4.0

The Bluetooth SIG completed the Bluetooth Core Specification version 4.0 (called Bluetooth Smart) and has been adopted as of 30 June 2010. It includes Classic Bluetooth,

Bluetooth high speed and Bluetooth Low Energy (BLE) protocols. Bluetooth high speed is based on Wi-Fi, and Classic Bluetooth consists of legacy Bluetooth protocols.

Bluetooth Low Energy, previously known as Wibree, is a subset of Bluetooth v4.0 with an entirely new protocol stack for rapid build-up of simple links. As an alternative to the Bluetooth standard protocols that were introduced in Bluetooth v1.0 to v3.0, it is aimed at very low power applications powered by a coin cell. Chip designs allow for two types of implementation, dual-mode, single-mode and enhanced past versions. The provisional names Wibree and Bluetooth ULP (Ultra Low Power) were abandoned and the BLE name was used for a while. In late 2011, new logos "Bluetooth Smart Ready" for hosts and "Bluetooth Smart" for sensors were introduced as the general-public face of BLE.

Compared to Classic Bluetooth, Bluetooth Low Energy is intended to provide considerably reduced power consumption and cost while maintaining a similar communication range. In terms of lengthening the battery life of Bluetooth devices, BLE represents a significant progression.

- In a single-mode implementation, only the low energy protocol stack is implemented. Dialog Semiconductor, STMicroelectronics, AMICCOM, CSR, Nordic Semiconductor and Texas Instruments have released single mode Bluetooth Low Energy solutions.

- In a dual-mode implementation, Bluetooth Smart functionality is integrated into an existing Classic Bluetooth controller. As of March 2011, the following semiconductor companies have announced the availability of chips meeting the standard: Qualcomm-Atheros, CSR, Broadcom and Texas Instruments. The compliant architecture shares all of Classic Bluetooth's existing radio and functionality resulting in a negligible cost increase compared to Classic Bluetooth.

Cost-reduced single-mode chips, which enable highly integrated and compact devices, feature a lightweight Link Layer providing ultra-low power idle mode operation, simple device discovery, and reliable point-to-multipoint data transfer with advanced power-save and secure encrypted connections at the lowest possible cost.

General improvements in version 4.0 include the changes necessary to facilitate BLE modes, as well the Generic Attribute Profile (GATT) and Security Manager (SM) services with AES Encryption.

- Core Specification Addendum 2 was unveiled in December 2011; it contains improvements to the audio Host Controller Interface and to the High Speed (802.11) Protocol Adaptation Layer.

- Core Specification Addendum 3 revision 2 has an adoption date of 24 July 2012.

- Core Specification Addendum 4 has an adoption date of 12 February 2013.

## Bluetooth 4.1

The Bluetooth SIG announced formal adoption of the Bluetooth v4.1 specification on 4 December 2013. This specification is an incremental software update to Bluetooth Specification v4.0, and not a hardware update. The update incorporates Bluetooth Core Specification Addenda (CSA 1, 2, 3 & 4) and adds new features that improve consumer usability. These include increased co-existence support for LTE, bulk data exchange rates—and aid developer innovation by allowing devices to support multiple roles simultaneously.

New features of this specification include:

- Mobile Wireless Service Coexistence Signaling.

- Train Nudging and Generalized Interlaced Scanning.

- Low Duty Cycle Directed Advertising.

- L2CAP Connection Oriented and Dedicated Channels with Credit-Based Flow Control.

- Dual Mode and Topology.

- LE Link Layer Topology.

- 802.11n PAL.

- Audio Architecture Updates for Wide Band Speech.

- Fast Data Advertising Interval.

- Limited Discovery Time.

Notice that some features were already available in a Core Specification Addendum (CSA) before the release of v4.1.

## Bluetooth 4.2

Released on December 2, 2014, it introduces features for the Internet of Things.

The major areas of improvement are:

- Low Energy Secure Connection with Data Packet Length Extension.

- Link Layer Privacy with Extended Scanner Filter Policies.

- Internet Protocol Support Profile (IPSP) version 6 ready for Bluetooth Smart things to support connected home.

Older Bluetooth hardware may receive 4.2 features such as Data Packet Length Extension and improved privacy via firmware updates.

## Bluetooth 5

The Bluetooth SIG presented Bluetooth 5 on 16 June 2016. Its new features are mainly focused on new Internet of Things technology. Sony was the first to announce Bluetooth 5.0 support with its Xperia XZ Premium in Feb 2017 during the Mobile World Congress 2017. The Samsung Galaxy S8 launched with Bluetooth 5 support in April 2017. In September 2017, the iPhone 8, 8 Plus and iPhone X launched with Bluetooth 5 support as well. Apple also integrated Bluetooth 5 in its new HomePod offering released on February 9, 2018. Marketing drops the point number; so that it is just "Bluetooth 5" (unlike Bluetooth 4.0). The change is for the sake of "Simplifying our marketing, communicating user benefits more effectively and making it easier to signal significant technology updates to the market."

Bluetooth 5 provides, for BLE, options that can double the speed (2 Mbit/s burst) at the expense of range, or up to fourfold the range at the expense of data rate. The increase in transmissions could be important for Internet of Things devices, where many nodes connect throughout a whole house. Bluetooth 5 adds functionality for connectionless services such as location-relevant navigation of low-energy Bluetooth connections.

The major areas of improvement are:

- Slot Availability Mask (SAM).

- 2 Mbit/s PHY for LE.

- LE Long Range.

- High Duty Cycle Non-Connectable Advertising.

- LE Advertising Extensions.

- LE Channel Selection Algorithm #2.

Features Added in CSA5 – Integrated in v5.0:

- Higher Output Power.

The following features were removed in this version of the specification:

- Park State.

## Bluetooth 5.1

The Bluetooth SIG presented Bluetooth 5.1 on 21 January 2019.

The major areas of improvement are:

- Angle of Arrival (AoA) and Angle of Departure (AoD) which are used for location and tracking of devices.
- Advertising Channel Index.
- GATT Caching.
- Minor Enhancements batch 1:
  - HCI support for debug keys in LE Secure Connections.
  - Sleep clock accuracy update mechanism.
  - ADI field in scan response data.
  - Interaction between QoS and Flow Specification.
  - Host channel classification for secondary advertising.
  - Allow the SID to appear in scan response reports.
  - Specify the behavior when rules are violated.
- Periodic Advertising Sync Transfer.

Features Added in Core Specification Addendum (CSA) 6 – Integrated in v5.1:

- Models.
- Mesh-based model hierarchy.

The following features were removed in this version of the specification:

- Unit keys.

## Bluetooth 5.2

In December 2019 the Bluetooth SIG published the Bluetooth Core Specification Version 5.2. The new specification adds three main new features:

- Enhanced Attribute Protocol (EATT), an improved version of the Attribute protocol (ATT).
- LE Power Control.
- LE Isochronous Channels.

## Architecture

### Software

Seeking to extend the compatibility of Bluetooth devices, the devices that adhere to the standard use as an interface between the host device (laptop, phone, etc.) and the Bluetooth device as such (Bluetooth chip) an interface called HCI (Host Controller Interface).

High-level protocols such as the SDP (Protocol used to find other Bluetooth devices within the communication range, also responsible for detecting the function of devices in range), RFCOMM (Protocol used to emulate serial port connections) and TCS (Telephony control protocol) interact with the baseband controller through the L2CAP Protocol (Logical Link Control and Adaptation Protocol). The L2CAP protocol is responsible for the segmentation and reassembly of the packets.

### Hardware

The hardware that makes up the Bluetooth device is made up of, logically, two parts; which may or may not be physically separate. A radio device, responsible for modulating and transmitting the signal; and a digital controller. The digital controller is likely a CPU, one of whose functions is to run a Link Controller; and interfaces with the host device; but some functions may be delegated to hardware. The Link Controller is responsible for the processing of the baseband and the management of ARQ and physical layer FEC protocols. In addition, it handles the transfer functions (both asynchronous and synchronous), audio coding and data encryption. The CPU of the device is responsible for attending the instructions related to Bluetooth of the host device, in order to simplify its operation. To do this, the CPU runs software called Link Manager that has the function of communicating with other devices through the LMP protocol.

A Bluetooth device is a short-range wireless device. Bluetooth devices are fabricated on RF CMOS integrated circuit (RF circuit) chips.

### Bluetooth Protocol Stack

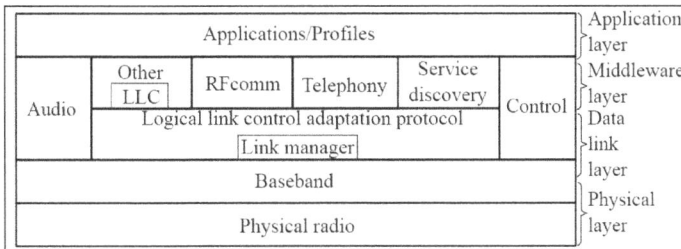

Bluetooth Protocol Stack.

Bluetooth is defined as a layer protocol architecture consisting of core protocols, cable replacement protocols, telephony control protocols, and adopted protocols. Mandatory protocols for all Bluetooth stacks are LMP, L2CAP and SDP. In addition, devices that communicate with Bluetooth almost universally can use these protocols: HCI and RF-COMM.

## Link Manager

The Link Manager (LM) is the system that manages establishing the connection between devices. It is responsible for the establishment, authentication and configuration of the link. The Link Manager locates other managers and communicates with them via the management protocol of the LMP link. In order to perform its function as a service provider, the LM uses the services included in the Link Controller (LC). The Link Manager Protocol basically consists of a number of PDUs (Protocol Data Units) that are sent from one device to another. The following is a list of supported services:

- Transmission and reception of data.

- Name request.

- Request of the link addresses.

- Establishment of the connection.

- Authentication.

- Negotiation of link mode and connection establishment.

## Host Controller Interface

The Host Controller Interface provides a command interface for the controller and for the link manager, which allows access to the hardware status and control registers. This interface provides an access layer for all Bluetooth devices. The HCI layer of the machine exchanges commands and data with the HCI firmware present in the Bluetooth device. One of the most important HCI tasks that must be performed is the automatic discovery of other Bluetooth devices that are within the coverage radius.

## Logical Link Control and Adaptation Protocol

The Logical Link Control and Adaptation Protocol (L2CAP) is used to multiplex multiple logical connections between two devices using different higher level protocols. Provides segmentation and reassembly of on-air packets.

In Basic mode, L2CAP provides packets with a payload configurable up to 64 kB, with 672 bytes as the default MTU, and 48 bytes as the minimum mandatory supported MTU.

In Retransmission and Flow Control modes, L2CAP can be configured either for isochronous data or reliable data per channel by performing retransmissions and CRC checks.

Bluetooth Core Specification Addendum 1 adds two additional L2CAP modes to the core specification. These modes effectively deprecate original Retransmission and Flow Control modes.

## Enhanced Retransmission Mode (ERTM)

This mode is an improved version of the original retransmission mode. This mode provides a reliable L2CAP channel.

## Streaming Mode (SM)

This is a very simple mode, with no retransmission or flow control. This mode provides an unreliable L2CAP channel.

Reliability in any of these modes is optionally and additionally guaranteed by the lower layer Bluetooth BDR/EDR air interface by configuring the number of retransmissions and flush timeout (time after which the radio flushes packets). In-order sequencing is guaranteed by the lower layer.

Only L2CAP channels configured in ERTM or SM may be operated over AMP logical links.

## Service Discovery Protocol

The Service Discovery Protocol (SDP) allows a device to discover services offered by other devices, and their associated parameters. For example, when you use a mobile phone with a Bluetooth headset, the phone uses SDP to determine which Bluetooth profiles the headset can use (Headset Profile, Hands Free Profile, Advanced Audio Distribution Profile (A2DP) etc.) and the protocol multiplexer settings needed for the phone to connect to the headset using each of them. Each service is identified by a Universally Unique Identifier (UUID), with official services (Bluetooth profiles) assigned a short form UUID (16 bits rather than the full 128).

## Radio Frequency Communications

Radio Frequency Communications (RFCOMM) is a cable replacement protocol used for generating a virtual serial data stream. RFCOMM provides for binary data transport and emulates EIA-232 (formerly RS-232) control signals over the Bluetooth baseband layer, i.e., it is a serial port emulation.

RFCOMM provides a simple, reliable, data stream to the user, similar to TCP. It is used directly by many telephony related profiles as a carrier for AT commands, as well as being a transport layer for OBEX over Bluetooth.

Many Bluetooth applications use RFCOMM because of its widespread support and publicly available API on most operating systems. Additionally, applications that used a serial port to communicate can be quickly ported to use RFCOMM.

## Bluetooth Network Encapsulation Protocol

The Bluetooth Network Encapsulation Protocol (BNEP) is used for transferring another protocol stack's data via an L2CAP channel. Its main purpose is the transmission of IP packets in the Personal Area Networking Profile. BNEP performs a similar function to SNAP in Wireless LAN.

## Audio/Video Control Transport Protocol

The Audio/Video Control Transport Protocol (AVCTP) is used by the remote control profile to transfer AV/C commands over an L2CAP channel. The music control buttons on a stereo headset use this protocol to control the music player.

## Audio/Video Distribution Transport Protocol

The Audio/Video Distribution Transport Protocol (AVDTP) is used by the advanced audio distribution (A2DP) profile to stream music to stereo headsets over an L2CAP channel intended for video distribution profile in the Bluetooth transmission.

## Telephony Control Protocol

The Telephony Control Protocol – Binary (TCS BIN) is the bit-oriented protocol that defines the call control signaling for the establishment of voice and data calls between Bluetooth devices. Additionally, "TCS BIN defines mobility management procedures for handling groups of Bluetooth TCS devices."

TCS-BIN is only used by the cordless telephony profile, which failed to attract implementers. As such it is only of historical interest.

## Adopted Protocols

Adopted protocols are defined by other standards-making organizations and incorporated into Bluetooth's protocol stack, allowing Bluetooth to code protocols only when necessary. The adopted protocols include:

- Point-to-Point Protocol (PPP): Internet standard protocol for transporting IP datagrams over a point-to-point link.

- TCP/IP/UDP: Foundation Protocols for TCP/IP protocol suite.

- Object Exchange Protocol (OBEX): Session-layer protocol for the exchange of objects, providing a model for object and operation representation.

- Wireless Application Environment/Wireless Application Protocol (WAE/WAP): WAE specifies an application framework for wireless devices and WAP is an open standard to provide mobile users access to telephony and information services.

## Baseband Error Correction

Depending on packet type, individual packets may be protected by error correction, either 1/3 rate forward error correction (FEC) or 2/3 rate. In addition, packets with CRC will be retransmitted until acknowledged by automatic repeat request (ARQ).

## Setting Up Connections

Any Bluetooth device in discoverable mode transmits the following information on demand:

- Device name.

- Device class.

- List of services.

- Technical information (for example: device features, manufacturer, Bluetooth specification used, clock offset).

Any device may perform an inquiry to find other devices to connect to, and any device can be configured to respond to such inquiries. However, if the device trying to connect knows the address of the device, it always responds to direct connection requests and transmits the information shown in the list above if requested. Use of a device's services may require pairing or acceptance by its owner, but the connection itself can be initiated by any device and held until it goes out of range. Some devices can be connected to only one device at a time, and connecting to them prevents them from connecting to other devices and appearing in inquiries until they disconnect from the other device.

Every device has a unique 48-bit address. However, these addresses are generally not shown in inquiries. Instead, friendly Bluetooth names are used, which can be set by the user. This name appears when another user scans for devices and in lists of paired devices.

Most cellular phones have the Bluetooth name set to the manufacturer and model of the phone by default. Most cellular phones and laptops show only the Bluetooth names and special programs are required to get additional information about remote devices. This can be confusing as, for example, there could be several cellular phones in range named T610.

## Pairing and Bonding

## Motivation

Many services offered over Bluetooth can expose private data or let a connecting party control the Bluetooth device. Security reasons make it necessary to recognize specific devices, and thus enable control over which devices can connect to a given Bluetooth device. At the same time, it is useful for Bluetooth devices to be able to establish a connection without user intervention (for example, as soon as in range).

To resolve this conflict, Bluetooth uses a process called bonding, and a bond is generated through a process called pairing. The pairing process is triggered either by a specific request from a user to generate a bond (for example, the user explicitly requests to "Add a Bluetooth device"), or it is triggered automatically when connecting to a service where (for the first time) the identity of a device is required for security purposes. These two cases are referred to as dedicated bonding and general bonding respectively.

Pairing often involves some level of user interaction. This user interaction confirms the identity of the devices. When pairing successfully completes, a bond forms between the two devices, enabling those two devices to connect to each other in the future without repeating the pairing process to confirm device identities. When desired, the user can remove the bonding relationship.

## Implementation

During pairing, the two devices establish a relationship by creating a shared secret known as a link key. If both devices store the same link key, they are said to be paired or bonded. A device that wants to communicate only with a bonded device can cryptographically authenticate the identity of the other device, ensuring it is the same device it previously paired with. Once a link key is generated, an authenticated Asynchronous Connection-Less (ACL) link between the devices may be encrypted to protect exchanged data against eavesdropping. Users can delete link keys from either device, which removes the bond between the devices—so it is possible for one device to have a stored link key for a device it is no longer paired with.

Bluetooth services generally require either encryption or authentication and as such require pairing before they let a remote device connect. Some services, such as the Object Push Profile, elect not to explicitly require authentication or encryption so that pairing does not interfere with the user experience associated with the service use-cases.

## Pairing mechanisms

Pairing mechanisms changed significantly with the introduction of Secure Simple Pairing in Bluetooth v2.1. The following summarizes the pairing mechanisms:

- Legacy pairing: This is the only method available in Bluetooth v2.0 and before. Each device must enter a PIN code; pairing is only successful if both devices enter the same PIN code. Any 16-byte UTF-8 string may be used as a PIN code; however, not all devices may be capable of entering all possible PIN codes.

- ◦ Limited input devices: The obvious example of this class of device is a Bluetooth Hands-free headset, which generally have few inputs. These devices usually have a fixed PIN, for example "0000" or "1234", that are hard-coded into the device.

- ◦ Numeric input devices: Mobile phones are classic examples of these devices. They allow a user to enter a numeric value up to 16 digits in length.

- ◦ Alpha-numeric input devices: PCs and smartphones are examples of these devices. They allow a user to enter full UTF-8 text as a PIN code. If pairing with a less capable device the user must be aware of the input limitations on the other device; there is no mechanism available for a capable device to determine how it should limit the available input a user may use.

- Secure Simple Pairing (SSP): This is required by Bluetooth v2.1, although a Bluetooth v2.1 device may only use legacy pairing to interoperate with a v2.0 or earlier device. Secure Simple Pairing uses a form of public key cryptography, and some types can help protect against man in the middle, or MITM attacks. SSP has the following authentication mechanisms:

- ◦ Just works: As the name implies, this method just works, with no user interaction. However, a device may prompt the user to confirm the pairing process. This method is typically used by headsets with very limited IO capabilities, and is more secure than the fixed PIN mechanism this limited set of devices uses for legacy pairing. This method provides no man-in-the-middle (MITM) protection.

- ◦ Numeric comparison: If both devices have a display, and at least one can accept a binary yes/no user input, they may use Numeric Comparison. This method displays a 6-digit numeric code on each device. The user should compare the numbers to ensure they are identical. If the comparison succeeds, the user(s) should confirm pairing on the device(s) that can accept an input. This method provides MITM protection, assuming the user confirms on both devices and actually performs the comparison properly.

- ◦ Passkey Entry: This method may be used between a device with a display and a device with numeric keypad entry (such as a keyboard), or two devices with numeric keypad entry. In the first case, the display presents a 6-digit numeric code to the user, who then enters the code on the keypad. In the second case, the user of each device enters the same 6-digit number. Both of these cases provide MITM protection.

- ◦ Out of band (OOB): This method uses an external means of communication, such as near-field communication (NFC) to exchange some information used in the pairing process. Pairing is completed using the Bluetooth radio, but requires information from the OOB mechanism. This provides only the level of MITM protection that is present in the OOB mechanism.

SSP is considered simple for the following reasons:

- In most cases, it does not require a user to generate a passkey.

- For use cases not requiring MITM protection, user interaction can be eliminated.

- For numeric comparison, MITM protection can be achieved with a simple equality comparison by the user.

- Using OOB with NFC enables pairing when devices simply get close, rather than requiring a lengthy discovery process.

## Security Concerns

Prior to Bluetooth v2.1, encryption is not required and can be turned off at any time. Moreover, the encryption key is only good for approximately 23.5 hours; using a single encryption key longer than this time allows simple XOR attacks to retrieve the encryption key.

- Turning off encryption is required for several normal operations, so it is problematic to detect if encryption is disabled for a valid reason or for a security attack.

Bluetooth v2.1 addresses this in the following ways:

- Encryption is required for all non-SDP (Service Discovery Protocol) connections.

- A new Encryption Pause and Resume feature is used for all normal operations that require that encryption be disabled. This enables easy identification of normal operation from security attacks.

- The encryption key must be refreshed before it expires.

Link keys may be stored on the device file system, not on the Bluetooth chip itself. Many Bluetooth chip manufacturers let link keys be stored on the device—however, if the device is removable, this means that the link key moves with the device.

## Security

Bluetooth implements confidentiality, authentication and key derivation with custom algorithms based on the SAFER+ block cipher. Bluetooth key generation is generally based on a Bluetooth PIN, which must be entered into both devices. This procedure might be modified if one of the devices has a fixed PIN (e.g., for headsets or similar devices with a restricted user interface). During pairing, an initialization key or master key is generated, using the E22 algorithm. The E0 stream cipher is used for encrypting packets, granting confidentiality, and is based on a shared cryptographic secret, namely a previously generated link key or master key. Those keys, used for subsequent encryption of data sent via the air interface, rely on the Bluetooth PIN, which has been entered into one or both devices.

An overview of Bluetooth vulnerabilities exploits was published in 2007 by Andreas Becker.

In September 2008, the National Institute of Standards and Technology (NIST) published a Guide to Bluetooth Security as a reference for organizations. It describes Bluetooth security capabilities and how to secure Bluetooth technologies effectively. While Bluetooth has its benefits, it is susceptible to denial-of-service attacks, eavesdropping, man-in-the-middle attacks, message modification, and resource misappropriation. Users and organizations must evaluate their acceptable level of risk and incorporate security into the lifecycle of Bluetooth devices. To help mitigate risks, included in the NIST document are security checklists with guidelines and recommendations for creating and maintaining secure Bluetooth piconets, headsets, and smart card readers.

Bluetooth v2.1 – finalized in 2007 with consumer devices first appearing in 2009 – makes significant changes to Bluetooth's security, including pairing.

### Bluejacking

Bluejacking is the sending of either a picture or a message from one user to an unsuspecting user through Bluetooth wireless technology. Common applications include short messages, e.g., "You've just been bluejacked!" Bluejacking does not involve the removal or alteration of any data from the device. Bluejacking can also involve taking control of a mobile device wirelessly and phoning a premium rate line, owned by the bluejacker. Security advances have alleviated this issue.

# NETWORK SERVICE PROVIDER

A network service provider (NSP) is a business or organization that sells bandwidth or network access by providing direct Internet backbone access to internet service providers and usually access to its network access points (NAPs). For such a reason, network service providers are sometimes referred to as *backbone providers* or *internet providers*.

Network service providers may consist of telecommunications companies, data carriers, wireless communications providers, Internet service providers, and cable television operators offering high-speed Internet access. It can also include certain information technology companies such as IBM, DXC Technology, Vanco and Atos Origin.

The primary customers of NSPs are other service providers, including internet service providers (ISPs), which, in turn, sell internet access to businesses and consumers. Several network service providers also function as ISPs themselves, however. NSPs are also referred to as backbone providers.

NSPs build and maintain the fiber optic cable and core routers - i.e., the principal data routes - that make up the internet. Their physical connections come together at internet exchange points, which is where regional ISPs can connect to an NSP backbone. These locations are also called peering points.

Not all network service providers are household names, but they are essential to modern networking. Their high-speed infrastructure and services support a downstream ecosystem that includes ISPs, wireless carriers and virtual network operators, among others, while also providing the foundation for all commercial IP services.

## Examples of NSPs

Each region of the globe typically has a handful of NSPs, although some regions have only one. Some of the world's top NSPs include the following companies:

- AT&T
- CenturyLink
- China Telecom
- Cogent
- Deutsche Telekom
- NTT
- Orange
- Sprint
- Tata
- Verizon Business

## Working of a Network Service Provider

An ISP can purchase wholesale bandwidth from an NSP, which provides connectivity for their customers. Customers then access the network through their ISP's last-mile infrastructure, which, in turn, connects to the NSP's backbone.

From there, the NSP routes all traffic and provides the infrastructure needed for network connectivity. The NSP builds, maintains and expands its infrastructure to meet traffic demands. The ISP is responsible for its own network, sales, marketing and customer service. ISPs can also purchase other services from an NSP - such as cloud-based services or web hosting - that are sold to the end customer under the ISP's brand name, a strategy known as white labeling.

# References

- Wireless-2993: techopedia.com, Retrieved 14 July, 2019

- Wireless-application-service-provider: searchmobilecomputing.techtarget.com, Retrieved 13 May, 2019

- IEEE Std 802.15.1–2002 – IEEE Standard for Information technology – Telecommunications and information exchange between systems – Local and metropolitan area networks – Specific requirements Part 15.1: Wireless Medium Access Control (MAC) and Physical Layer (PHY) Specifications for Wireless Personal Area Networks (WPANs). Ieeexplore.ieee.org. 2002. doi:10.1109/IEEESTD.2002.93621. ISBN 978-0-7381-3335-5. Retrieved 4 September 2010

- Wireless-internet-service-provider-wisp-2546: techopedia.com, Retrieved 27 March, 2019

- Guowang Miao, Jens Zander, Ki Won Sung, and Ben Slimane, Fundamentals of Mobile Data Networks, Cambridge University Press, ISBN 1107143217, 2016

- Network-service-provider: searchnetworking.techtarget.com, Retrieved 24 February, 2019

# Mobile WiMAX

Mobile WiMAX is a technical wireless system that enables convergence of mobile and fixed broadband networks through a common wide area network. Mobile IP, 3G IEEE family, 4G IEEE family, access service network gateway, etc. are some of the concepts that fall under mobile WiMAX. All these concepts of mobile WiMAX have been carefully analyzed in this chapter.

## MOBILE BROADBAND WIRELESS ACCESS TECHNOLOGIES

Wireless broadband technologies include services from mobile phone service providers which allow a more mobile version of Internet access. Consumers can purchase a PC card, laptop card, or USB equipment to connect their PC or laptop to the Internet via cell phone towers. This type of connection would be stable in almost any area that could also receive a strong cell phone connection. These connections can cost more for portable convenience as well as having speed limitations in all but urban environments.

Emerging telecommunications applications such as multimedia streaming, music download, online gaming, and content browsing are popular examples of the digital revolution we have been facing, as the world gets connected. Fixed broadband access has already become an urban commodity in the developed countries, but so far there have been few means of delivering these bandwidth-consuming services effectively and affordably to the significant number of rural and mobile users. However, recent advances in e.g. signal processing, radio protocols, and mobile network infrastructure are now enabling the concept of mobile broadband for consumers around the world.

Mobile broadband is defined here as the potential to transfer low-latency user data with speeds exceeding 256 kbit/s while roaming the network with seamless handovers between adjacent cells.

### Mobile Broadband Technologies

Currently there are a number of different technologies for broadband wireless access (BWA) for both fixed and mobile applications. Some of them are completely proprietary, based on vendor-specific solutions that are non-interoperable, while others are based on open standards developed by industry working groups.

## Mobile WiMAX (IEEE 802.16e-2005)

Perhaps the biggest shortcoming of 802.16-2004 is the lack of support for mobility. IEEE addressed this issue by developing specifications for a separate version of the standard, the 802.16e, which was approved on December 7, 2005 (IEEE 2005). Also known as mobile WiMAX, the standard is seen to be in competition with 3G cellular technologies. Its radio access method is even more sophisticated than that of fixed WiMAX, utilizing scalable OFDMA and thus achieving an even better link budget. The tradeoff is increased complexity in physical layer processing. Fast handover signaling is supported, e.g., to allow users in moving vehicles to seamlessly switch between base stations.

Mobile WiMAX operates in the 2 to 6 GHz range that mainly consists of licensed bands. Mobile applications are likely to operate in frequencies below 3 GHz, while even some fixed applications are expected to use 802.16e due to its better characteristics. However, it should be noted that there is no backward compatibility with fixed WiMAX. Cell radiuses are expected to be typically 2 to 5 km, and user data rates up to 30 Mbit/s are achievable in theory with full 10 MHz channels. The first certified 802.16e products are expected to be available by late 2006, though wide scale commercial deployments are expected not earlier than 2008.

## MBWA (IEEE 802.20)

The IEEE 802.20 (or Mobile Broadband Wireless Access) Working Group was established on December 11, 2002 with the aim to develop a specification for an efficient packet based air interface that is optimized for the transport of IP based services. The goal is to enable worldwide deployment of affordable, always-on, and interoperable BWA networks for both business and residential end user markets. The group will specify the lower layers of the air interface, operating in licensed bands below 3.5 GHz and enabling peak user data rates exceeding 1 Mbit/s at speeds of up to 250 km/h. (IEEE 2006a).

The goals of 802.20 and 802.16e are similar. However, 802.16e is much more mature, whereas even the standardization process of 802.20 is far from complete. A draft version of the specification was, however, approved on January 18, 2006 (IEEE 2006b).

## Flash-OFDM

Flash-OFDM, short for Fast Low-latency Access with Seamless Handoff OFDM, is a proprietary wireless broadband technology originally developed by Flarion Technologies which was recently acquired by Qualcomm, a major developer and patent holder of Code Division Multiple Access (CDMA) and other advanced wireless technologies.

As the name implies, Flash-OFDM's radio access method utilizes OFDM in relatively narrow 1.25 MHz FDD channels. Frequency hopping is employed in the subcarriers, which provides frequency diversity. Operation is supported in several licensed frequency bands, such as 450 MHz, 700 MHz, 800 MHz, 1.9 GHz, and 2.1 GHz. The network is all-IP based, and inherently supports applications such as VoIP due to its low latency and enhanced QoS support. Flash-OFDM is claimed to reach user data rates of 1 to 1.5 Mbit/s in downlink and around 300 to 500 kbit/s in uplink, with a typical latency of 50 ms.

Compared to mobile WiMAX, Flash-OFDM has a time-to-market advantage in that its equipment is readily available on the market, but a major disadvantage in having only limited vendor support and not being an open technology. Interestingly, Flash-OFDM is also a candidate for the IEEE 802.20 standardization effort.

# MOBILE IP

Mobile IP (or MIP) is an Internet Engineering Task Force (IETF) standard communications protocol that is designed to allow mobile device users to move from one network to another while maintaining a permanent IP address. Mobile IP for IPv4 is described in IETF RFC 5944, and extensions are defined in IETF RFC 4721. Mobile IPv6, the IP mobility implementation for the next generation of the Internet Protocol, IPv6, is described in RFC 6275.

The Mobile IP allows for location-independent routing of IP datagrams on the Internet. Each mobile node is identified by its home address disregarding its current location in the Internet. While away from its home network, a mobile node is associated with a care-of address which identifies its current location and its home address is associated with the local endpoint of a tunnel to its home agent. Mobile IP specifies how a mobile node registers with its home agent and how the home agent routes datagrams to the mobile node through the tunnel.

## Applications

In many applications (e.g., VPN, VoIP), sudden changes in network connectivity and IP address can cause problems. Mobile IP was designed to support seamless and continuous Internet connectivity.

Mobile IP is most often found in wired and wireless environments where users need to carry their mobile devices across multiple LAN subnets. Examples of use are in roaming between overlapping wireless systems, e.g., IP over DVB, WLAN, WiMAX and BWA.

Mobile IP is not required within cellular systems such as 3G, to provide transparency when Internet users migrate between cellular towers, since these systems provide their own data link layer handover and roaming mechanisms. However, it is often used in 3G systems to allow seamless IP mobility between different packet data serving node (PDSN) domains.

## Operational Principles

The goal of IP Mobility is to maintain the TCP connection between a mobile host and a static host while reducing the effects of location changes while the mobile host is moving around, without having to change the underlying TCP/IP. To solve the problem, the RFC allows for a kind of proxy agent to act as a *middle-man* between a mobile host and a correspondent host.

A mobile node has two addresses – a permanent home address and a care-of address (CoA), which is associated with the network the mobile node is visiting. Two kinds of entities comprise a Mobile IP implementation:

- A home agent (HA) stores information about mobile nodes whose permanent home address is in the home agent's network. The HA acts as a router on a mobile host's (MH) home network which tunnels datagrams for delivery to the MH when it is away from home, maintains a location directory (LD) for the MH.

- A foreign agent (FA) stores information about mobile nodes visiting its network. Foreign agents also advertise care-of-addresses, which are used by Mobile IP. If there is no foreign agent in the host network, the mobile device has to take care of getting an address and advertising that address by its own means. The FA acts as a router on a MH's visited network which provides routing services to the MH while registered. FA detunnels and delivers datagrams to the MH that were tunneled by the MH's HA.

The so-called Care of Address is a termination point of a tunnel toward a MH, for datagrams forwarded to the MH while it is away from home.

- Foreign agent care-of address: The address of a foreign agent that MH registers with.

- Co-located care-of address: An externally obtained local address that a MH gets.

A Mobile Node (MN) is responsible for discovering whether it is connected to its home network or has moved to a foreign network. HA's and FA's broadcast their presence on each network to which they are attached. They are not solely *responsible* for discovery, they only play a part. RFC 2002 specified that MN use agent discovery to locate these entities. When connected to a foreign network, a MN has to determine the foreign agent care-of-address being offered by each foreign agent on the network.

A node wanting to communicate with the mobile node uses the permanent home address of the mobile node as the destination address to send packets to. Because the home address logically belongs to the network associated with the home agent, normal IP routing mechanisms forward these packets to the home agent. Instead of forwarding these packets to a destination that is physically in the same network as the home agent, the home agent redirects these packets towards the remote address through an IP tunnel by encapsulating the datagram with a new IP header using the care of address of the mobile node.

When acting as transmitter, a mobile node sends packets directly to the other communicating node, without sending the packets through the home agent, using its permanent home address as the source address for the IP packets. This is known as triangular routing or "route optimization" (RO) mode. If needed, the foreign agent could employ reverse tunneling by tunneling the mobile node's packets to the home agent, which in turn forwards them to the communicating node. This is needed in networks whose gateway routers check that the source IP address of the mobile host belongs to their subnet or discard the packet otherwise. In Mobile IPv6 (MIPv6), "reverse tunneling" is the default behaviour, with RO being an optional behaviour.

## Development

Enhancements to the Mobile IP technique, such as Mobile IPv6 and Hierarchical Mobile IPv6 (HMIPv6) are being developed to improve mobile communications in certain circumstances by making the processes more secure and more efficient.

Researchers create support for mobile networking without requiring any pre-deployed infrastructure as it currently is required by MIP. One such example is Interactive Protocol for Mobile Networking (IPMN) which promises supporting mobility on a regular IP network just from the network edges by intelligent signalling between IP at end-points and application layer module with improved quality of service.

Researchers are also working to create support for mobile networking between entire subnets with support from Mobile IPv6. One such example is Network Mobility (NEMO) Network Mobility Basic Support Protocol by the IETF Network Mobility Working Group which supports mobility for entire Mobile Networks that move and to attach to different points in the Internet. The protocol is an extension of Mobile IPv6 and allows session continuity for every node in the Mobile Network as the network moves.

## Changes in IPv6 for Mobile IPv6

- A set of mobility options to include in mobility messages.

- A new Home Address option for the Destination Options header.

- A new Type 2 Routing header.

- New Internet Control Message Protocol for IPv6 (ICMPv6) messages to discover the set of home agents and to obtain the prefix of the home link.

- Changes to router discovery messages and options and additional Neighbor Discovery options.

- Foreign Agents are no longer needed.

## Important Terms

- Home network: The home network of a mobile device is the network within which the device receives its identifying IP address (home address).

- Home address: The home address of a mobile device is the IP address assigned to the device within its home network.

- Foreign network: A foreign network is the network in which a mobile node is operating when away from its home network.

- Care-of address: The care-of address of a mobile device is the network-native IP address of the device when operating in a foreign network.

- Home agent: A home agent is a router on a mobile node's home network which tunnels datagrams for delivery to the mobile node when it is away from home. It maintains current location (IP address) information for the mobile node. It is used with one or more foreign agents.

- Foreign agent: A foreign agent is a router that stores information about mobile nodes visiting its network. Foreign agents also advertise care-of-addresses which are used by Mobile IP.

- Binding: A binding is the association of the home address with a care-of address.

# 3G IEEE FAMILY

3G is the third generation of wireless mobile telecommunications technology. It is the upgrade for 2G and 2.5G GPRS networks, for faster data transfer speed. This is based on a set of standards used for mobile devices and mobile telecommunications use services and networks that comply with the International Mobile Telecommunications-2000 (IMT-2000) specifications by the International Telecommunication Union. 3G finds application in wireless voice telephony, mobile Internet access, fixed wireless Internet access, video calls and mobile TV.

3G telecommunication networks support services that provide an information transfer rate of at least 144 kbit/s. Later 3G releases, often denoted 3.5G and 3.75G, also provide mobile broadband access of several Mbit/s to smartphones and mobile modems in laptop computers. This ensures it can be applied to wireless voice telephony, mobile Internet access, fixed wireless Internet access, video calls and mobile TV technologies.

A new generation of cellular standards has appeared approximately every tenth year since 1G systems were introduced in 1979 and the early to mid-1980s. Each generation is characterized by new frequency bands, higher data rates and non–backward-compatible transmission technology. The first commercial 3G networks were introduced in 2000.

Several telecommunications companies market wireless mobile Internet services as 3G, indicating that the advertised service is provided over a 3G wireless network. Services advertised as 3G are required to meet IMT-2000 technical standards, including standards for reliability and speed (data transfer rates). To meet the IMT-2000 standards, a system is required to provide peak data rates of at least 144 kbit/s. However, many services advertised as 3G provide higher speed than the minimum technical requirements for a 3G service. Recent 3G releases, often denoted 3.5G and 3.75G, also provide mobile broadband access of several Mbit/s to smartphones and mobile modems in laptop computers.

The following standards are typically branded 3G:

- The UMTS (Universal Mobile Telecommunications System) system, first offered in 2001, standardized by 3GPP, used primarily in Europe, Japan, China (however with a different radio interface) and other regions predominated by GSM (Global Systems for Mobile) 2G system infrastructure. The cell phones are typically UMTS and GSM hybrids. Several radio interfaces are offered, sharing the same infrastructure:

  ○ The original and most widespread radio interface is called W-CDMA (Wideband Code Division Multiple Access).

  ○ The TD-SCDMA radio interface was commercialized in 2009 and is only offered in China.

  ○ The latest UMTS release, HSPA+, can provide peak data rates up to 56 Mbit/s in the downlink in theory (28 Mbit/s in existing services) and 22 Mbit/s in the uplink.

- The CDMA2000 system, first offered in 2002, standardized by 3GPP2, used especially in North America and South Korea, sharing infrastructure with the IS-95 2G standard. The cell phones are typically CDMA2000 and IS-95 hybrids. The latest release EVDO Rev B offers peak rates of 14.7 Mbit/s downstream.

The above systems and radio interfaces are based on spread spectrum radio transmission technology. While the GSM EDGE standard ("2.9G"), DECT cordless phones and Mobile WiMAX standards formally also fulfill the IMT-2000 requirements and are approved as 3G standards by ITU, these are typically not branded 3G, and are based on completely different technologies.

The following common standards comply with the IMT2000/3G standard:

- EDGE, a revision by the 3GPP organization to the older 2G GSM based transmission methods, utilizing the same switching nodes, base station sites and frequencies as GPRS, but new base station and cellphone RF circuits. It is based on the three times as efficient 8PSK modulation scheme as supplement to the original GMSK modulation scheme. EDGE is still used extensively due to its ease of upgrade from existing 2G GSM infrastructure and cell-phones.

  ◦ EDGE combined with the GPRS 2.5G technology is called EGPRS, and allows peak data rates in the order of 200 kbit/s, just as the original UMTS WCDMA versions, and thus formally fulfills the IMT2000 requirements on 3G systems. However, in practice EDGE is seldom marketed as a 3G system, but a 2.9G system. EDGE shows slightly better system spectral efficiency than the original UMTS and CDMA2000 systems, but it is difficult to reach much higher peak data rates due to the limited GSM spectral bandwidth of 200 kHz, and it is thus a dead end.

  ◦ EDGE was also a mode in the IS-136 TDMA system, today ceased.

  ◦ Evolved EDGE, the latest revision, has peaks of 1 Mbit/s downstream and 400 kbit/s upstream, but is not commercially used.

- The Universal Mobile Telecommunications System, created and revised by the 3GPP. The family is a full revision from GSM in terms of encoding methods and hardware, although some GSM sites can be retrofitted to broadcast in the UMTS/W-CDMA format.

  ◦ W-CDMA is the most common deployment, commonly operated on the 2,100 MHz band. A few others use the 10, 900 and 1,900 MHz bands.

    ▪ HSPA is an amalgamation of several upgrades to the original W-CDMA standard and offers speeds of 14.4 Mbit/s down and 5.76 Mbit/s up. HSPA is backward-compatible with and uses the same frequencies as W-CDMA.

    ▪ HSPA+, a further revision and upgrade of HSPA, can provide theoretical peak data rates up to 168 Mbit/s in the downlink and 22 Mbit/s in the uplink, using a combination of air interface improvements as well as multi-carrier HSPA and MIMO. Technically though, MIMO and DC-HSPA can be used without the "+" enhancements of HSPA+.

- The CDMA2000 system, or IS-2000, including CDMA2000 1x and CDMA2000 High Rate Packet Data (or EVDO), standardized by 3GPP2 (*differing* from the

3GPP), evolving from the original IS-95 CDMA system, is used especially in North America, China, India, Pakistan, Japan, South Korea, Southeast Asia, Europe and Africa.

- ○ CDMA2000 1x Rev. E has an increased voice capacity (in excess of three times) compared to Rev. 0 EVDO Rev. B offers downstream peak rates of 14.7 Mbit/s while Rev. C enhanced existing and new terminal user experience.

While DECT cordless phones and Mobile WiMAX standards formally also fulfill the IMT-2000 requirements, they are not usually considered due to their rarity and unsuitability for usage with mobile phones.

## Break-up of 3G systems

The 3G (UMTS and CDMA2000) research and development projects started in 1992. In 1999, ITU approved five radio interfaces for IMT-2000 as a part of the ITU-R M.1457 Recommendation; WiMAX was added in 2007.

There are evolutionary standards (EDGE and CDMA) that are backward-compatible extensions to pre-existing 2G networks as well as revolutionary standards that require all-new network hardware and frequency allocations. The cell phones use UMTS in combination with 2G GSM standards and bandwidths, but do not support EDGE. The latter group is the UMTS family, which consists of standards developed for IMT-2000, as well as the independently developed standards DECT and WiMAX, which were included because they fit the IMT-2000 definition.

While EDGE fulfills the 3G specifications, most GSM/UMTS phones report EDGE ("2.75G") and UMTS ("3G") functionality.

## Features

## Data Rates

ITU has not provided a clear definition of the data rate that users can expect from 3G equipment or providers. Thus users sold 3G service may not be able to point to a standard and say that the rates it specifies are not being met. While stating in commentary that "it is expected that IMT-2000 will provide higher transmission rates: a minimum data rate of 2 Mbit/s for stationary or walking users, and 348 kbit/s in a moving vehicle," the ITU does not actually clearly specify minimum required rates, nor required average rates, nor what modes of the interfaces qualify as 3G, so various data rates are sold as '3G' in the market.

In market implementation, 3G downlink data speeds defined by telecom service providers vary depending on the underlying technology deployed; up to 384kbit/s for WCDMA, up to 7.2Mbit/sec for HSPA and a theoretical maximum of 21.6 Mbit/s for HSPA+ (technically 3.5G, but usually clubbed under the tradename of 3G).

## Security

3G networks offer greater security than their 2G predecessors. By allowing the UE (User Equipment) to authenticate the network it is attaching to, the user can be sure the network is the intended one and not an impersonator. 3G networks use the KASUMI block cipher instead of the older A5/1 stream cipher. However, a number of serious weaknesses in the KASUMI cipher have been identified.

In addition to the 3G network infrastructure security, end-to-end security is offered when application frameworks such as IMS are accessed, although this is not strictly a 3G property.

## Applications of 3G

The bandwidth and location information available to 3G devices gives rise to applications not previously available to mobile phone users.

## Evolution

Both 3GPP and 3GPP2 are working on the extensions to 3G standards that are based on an all-IP network infrastructure and using advanced wireless technologies such as MIMO. These specifications already display features characteristic for IMT-Advanced (4G), the successor of 3G. However, falling short of the bandwidth requirements for 4G (which is 1 Gbit/s for stationary and 100 Mbit/s for mobile operation), these standards are classified as 3.9G or Pre-4G.

3GPP plans to meet the 4G goals with LTE Advanced, whereas Qualcomm has halted development of UMB in favour of the LTE family.

# 4G IEEE FAMILY

4G is the fourth generation of broadband cellular network technology, succeeding 3G. A 4G system must provide capabilities defined by ITU in IMT Advanced. Potential and current applications include amended mobile web access, IP telephony, gaming services, high-definition mobile TV, video conferencing, and 3D television.

The first-release Long Term Evolution (LTE) standard was commercially deployed in Oslo, Norway, and Stockholm, Sweden in 2009, and has since been deployed throughout most parts of the world. It has, however, been debated whether first-release versions should be considered 4G LTE.

In March 2009, the International Telecommunications Union-Radio communications sector (ITU-R) specified a set of requirements for 4G standards, named the International

Mobile Telecommunications Advanced (IMT-Advanced) specification, setting peak speed requirements for 4G service at 100 megabits per second (Mbit/s)(=12.5 megabytes per second) for high mobility communication (such as from trains and cars) and 1 gigabit per second (Gbit/s) for low mobility communication (such as pedestrians and stationary users).

Since the first-release versions of Mobile WiMAX and LTE support much less than 1 Gbit/s peak bit rate, they are not fully IMT-Advanced compliant, but are often branded 4G by service providers. According to operators, a generation of the network refers to the deployment of a new non-backward-compatible technology. On December 6, 2010, ITU-R recognized that these two technologies, as well as other beyond-3G technologies that do not fulfill the IMT-Advanced requirements, could nevertheless be considered "4G", provided they represent forerunners to IMT-Advanced compliant versions and "a substantial level of improvement in performance and capabilities with respect to the initial third generation systems now deployed".

Mobile WiMAX Release 2 (also known as WirelessMAN-Advanced or IEEE 802.16m') and LTE Advanced (LTE-A) are IMT-Advanced compliant backwards compatible versions of the above two systems, standardized during the spring 2011, and promising speeds in the order of 1 Gbit/s. Services were expected in 2013.

As opposed to earlier generations, a 4G system does not support traditional circuit-switched telephony service, but instead relies on all-Internet Protocol (IP) based communication such as IP telephony. As seen below, the spread spectrum radio technology used in 3G systems is abandoned in all 4G candidate systems and replaced by OFDMA multi-carrier transmission and other frequency-domain equalization (FDE) schemes, making it possible to transfer very high bit rates despite extensive multipath radio propagation (echoes). The peak bit rate is further improved by smart antenna arrays for multiple-input multiple-output (MIMO) communications.

In the field of mobile communications, a "generation" generally refers to a change in the fundamental nature of the service, non-backwards-compatible transmission technology, higher peak bit rates, new frequency bands, wider channel frequency bandwidth in Hertz, and higher capacity for many simultaneous data transfers (higher system spectral efficiency in bit/second/Hertz/site).

New mobile generations have appeared about every ten years since the first move from 1981 analog (1G) to digital (2G) transmission in 1992. This was followed, in 2001, by 3G multi-media support, spread spectrum transmission and, at least, 200 kbit/s peak bit rate, in 2011/2012 to be followed by "real" 4G, which refers to all-Internet Protocol (IP) packet-switched networks giving mobile ultra-broadband (gigabit speed) access.

While the ITU has adopted recommendations for technologies that would be used for future global communications, they do not actually perform the standardization or development work themselves, instead relying on the work of other standard bodies such as IEEE, WiMAX Forum, and 3GPP.

In the mid-1990s, the ITU-R standardization organization released the IMT-2000 requirements as a framework for what standards should be considered 3G systems, requiring 200 kbit/s peak bit rate. In 2008, ITU-R specified the IMT Advanced (International Mobile Telecommunications Advanced) requirements for 4G systems.

The fastest 3G-based standard in the UMTS family is the HSPA+ standard, which is commercially available since 2009 and offers 28 Mbit/s downstream (22 Mbit/s upstream) without MIMO, i.e. only with one antenna, and in 2011 accelerated up to 42 Mbit/s peak bit rate downstream using either DC-HSPA+ (simultaneous use of two 5 MHz UMTS carriers) or 2x2 MIMO. In theory speeds up to 672 Mbit/s are possible, but have not been deployed yet. The fastest 3G-based standard in the CDMA2000 family is the EV-DO Rev. B, which is available since 2010 and offers 15.67 Mbit/s downstream.

## IMT-advanced Requirements

Here, we refer to 4G using IMT-Advanced (International Mobile Telecommunications Advanced), as defined by ITU-R. An IMT-Advanced cellular system must fulfill the following requirements:

- Be based on an all-IP packet switched network.

- Have peak data rates of up to approximately 100 Mbit/s for high mobility such as mobile access and up to approximately 1 Gbit/s for low mobility such as nomadic/local wireless access.

- Be able to dynamically share and use the network resources to support more simultaneous users per cell.

- Use scalable channel bandwidths of 5–20 MHz, optionally up to 40 MHz.

- Have peak link spectral efficiency of 15 bit/s·Hz in the downlink, and 6.75 bit/s·Hz in the up link (meaning that 1 Gbit/s in the downlink should be possible over less than 67 MHz bandwidth).

- System spectral efficiency is, in indoor cases, 3 bit/s·Hz·cell for downlink and 2.25 bit/s·Hz·cell for up link.

- Smooth handovers across heterogeneous networks.

In September 2009, the technology proposals were submitted to the International Telecommunication Union (ITU) as 4G candidates. Basically all proposals are based on two technologies:

- LTE Advanced standardized by the 3GPP.

- 802.16m standardized by the IEEE.

Implementations of Mobile WiMAX and first-release LTE were largely considered a stopgap solution that would offer a considerable boost until WiMAX 2 (based on the 802.16m specification) and LTE Advanced was deployed. The latter's standard versions were ratified in spring 2011.

The first set of 3GPP requirements on LTE Advanced was approved in June 2008. LTE Advanced was standardized in 2010 as part of Release 10 of the 3GPP specification.

Some sources consider first-release LTE and Mobile WiMAX implementations as pre-4G or near-4G, as they do not fully comply with the planned requirements of 1 Gbit/s for stationary reception and 100 Mbit/s for mobile.

Confusion has been caused by some mobile carriers who have launched products advertised as 4G but which according to some sources are pre-4G versions, commonly referred to as 3.9G, which do not follow the ITU-R defined principles for 4G standards, but today can be called 4G according to ITU-R. Vodafone Netherlands for example, advertised LTE as 4G, while advertising LTE Advanced as their '4G+' service. A common argument for branding 3.9G systems as new-generation is that they use different frequency bands from 3G technologies; that they are based on a new radio-interface paradigm; and that the standards are not backwards compatible with 3G, whilst some of the standards are forwards compatible with IMT-2000 compliant versions of the same standards.

## System Standards

## IMT-2000 Compliant 4G Standards

As of October 2010, ITU-R Working Party 5D approved two industry-developed technologies (LTE Advanced and WirelessMAN-Advanced) for inclusion in the ITU's International Mobile Telecommunications Advanced program (IMT-Advanced program), which is focused on global communication systems that will be available several years from now.

## LTE Advanced

LTE Advanced (Long Term Evolution Advanced) is a candidate for IMT-Advanced standard, formally submitted by the 4GPP organization to ITU-T in the fall 2009, and expected to be released in 2013. The target of 3GPP LTE Advanced is to reach and surpass the ITU requirements. LTE Advanced is essentially an enhancement to LTE. It is not a new technology, but rather an improvement on the existing LTE network. This upgrade path makes it more cost effective for vendors to offer LTE and then upgrade to LTE Advanced which is similar to the upgrade from WCDMA to HSPA. LTE and LTE Advanced will also make use of additional spectrums and multiplexing to allow it to achieve higher data speeds. Coordinated Multi-point Transmission will also allow more system capacity to help handle the enhanced data speeds. Release 10 of LTE is expected to achieve the IMT Advanced speeds. Release 8 currently supports up to 300 Mbit/s of download speeds which is still short of the IMT-Advanced standards.

| Data speeds of LTE-Advanced | |
|---|---|
| | LTE Advanced |
| Peak download | 1000 Mbit/s |
| Peak upload | 0500 Mbit/s |

## IEEE 802.16m or WirelessMAN-Advanced

The IEEE 802.16m or WirelessMAN-Advanced evolution of 802.16e is under development, with the objective to fulfill the IMT-Advanced criteria of 1 Gbit/s for stationary reception and 100 Mbit/s for mobile reception.

## Forerunner Versions

## 3GPP Long Term Evolution (LTE)

Telia-branded LTE modem.

The pre-4G 3GPP Long Term Evolution (LTE) technology is often branded "4G – LTE", but the first LTE release does not fully comply with the IMT-Advanced requirements. LTE has a theoretical net bit rate capacity of up to 100 Mbit/s in the downlink and 50 Mbit/s in the uplink if a 20 MHz channel is used — and more if multiple-input multiple-output (MIMO), i.e. antenna arrays, are used.

4G+ Dual Band Modem.

The physical radio interface was at an early stage named High Speed OFDM Packet Access (HSOPA), now named Evolved UMTS Terrestrial Radio Access (E-UTRA). The first LTE USB dongles do not support any other radio interface.

The world's first publicly available LTE service was opened in the two Scandinavian capitals, Stockholm (Ericsson and Nokia Siemens Networks systems) and Oslo (a Huawei system) on December 14, 2009, and branded 4G. The user terminals were manufactured by Samsung. As of November 2012, the five publicly available LTE services in the United States are provided by MetroPCS, Verizon Wireless, AT&T Mobility, U.S. Cellular, Sprint, and T-Mobile US.

T-Mobile Hungary launched a public beta test (called friendly user test) on 7 October 2011, and has offered commercial 4G LTE services since 1 January 2012.

In South Korea, SK Telecom and LG U+ have enabled access to LTE service since 1 July 2011 for data devices, slated to go nationwide by 2012. KT Telecom closed its 2G service by March 2012, and complete the nationwide LTE service in the same frequency around 1.8 GHz by June 2012.

In the United Kingdom, LTE services were launched by EE in October 2012, by O2 and Vodafone in August 2013, and by Three in December 2013.

| Data speeds of LTE | |
|---|---|
| | LTE |
| Peak download | 0100 Mbit/s |
| Peak upload | 0050 Mbit/s |

## Mobile WiMAX (IEEE 802.16e)

The Mobile WiMAX (IEEE 802.16e-2005) mobile wireless broadband access (MWBA) standard (also known as WiBro in South Korea) is sometimes branded 4G, and offers peak data rates of 128 Mbit/s downlink and 56 Mbit/s uplink over 20 MHz wide channels.

In June 2006, the world's first commercial mobile WiMAX service was opened by KT in Seoul, South Korea.

Sprint has begun using Mobile WiMAX, as of 29 September 2008, branding it as a "4G" network even though the current version does not fulfill the IMT Advanced requirements on 4G systems.

In Russia, Belarus and Nicaragua WiMax broadband internet access were offered by a Russian company Scartel, and was also branded 4G, Yota.

| Data speeds of WiMAX | |
|---|---|
| | WiMAX |
| Peak download | 0128 Mbit/s |
| Peak upload | 0056 Mbit/s |

In the latest version of the standard, WiMax 2.1, the standard has been updated to be not compatible with earlier WiMax standard, and is instead interchangeable with LTE-TDD system, effectively merging WiMax standard with LTE.

## TD-LTE for China Market

Just as Long-Term Evolution (LTE) and WiMAX are being vigorously promoted in the global telecommunications industry, the former (LTE) is also the most powerful 4G mobile communications leading technology and has quickly occupied the Chinese market. TD-LTE, one of the two variants of the LTE air interface technologies, is not yet mature, but many domestic and international wireless carriers are, one after the other turning to TD-LTE.

IBM's data shows that 67% of the operators are considering LTE because this is the main source of their future market. The above news also confirms IBM's statement that while only 8% of the operators are considering the use of WiMAX, WiMAX can provide the fastest network transmission to its customers on the market and could challenge LTE.

TD-LTE is not the first 4G wireless mobile broadband network data standard, but it is China's 4G standard that was amended and published by China's largest telecom operator – China Mobile. After a series of field trials, is expected to be released into the commercial phase in the next two years. Ulf Ewaldsson, Ericsson's vice president said: "the Chinese Ministry of Industry and China Mobile in the fourth quarter of this year will hold a large-scale field test, by then, Ericsson will help the hand." But viewing from the current development trend, whether this standard advocated by China Mobile will be widely recognized by the international market is still debatable.

## Discontinued Candidate Systems

### UMB

UMB (Ultra Mobile Broadband) was the brand name for a discontinued 4G project within the 3GPP2 standardization group to improve the CDMA2000 mobile phone standard for next generation applications and requirements. In November 2008, Qualcomm, UMB's lead sponsor, announced it was ending development of the technology, favouring LTE instead. The objective was to achieve data speeds over 275 Mbit/s downstream and over 75 Mbit/s upstream.

### Flash-OFDM

At an early stage the Flash-OFDM system was expected to be further developed into a 4G standard.

### iBurst and MBWA (IEEE 802.20) Systems

The iBurst system (or HC-SDMA, High Capacity Spatial Division Multiple Access) was at an early stage considered to be a 4G predecessor. It was later further developed into the Mobile Broadband Wireless Access (MBWA) system, also known as IEEE 802.20.

## Principal Technologies in all Candidate Systems

### Features

The following key features can be observed in all suggested 4G technologies:

- Physical layer transmission techniques are as follows:
    - MIMO: To attain ultra high spectral efficiency by means of spatial processing including multi-antenna and multi-user MIMO.
    - Frequency-domain-equalization, for example multi-carrier modulation (OFDM) in the downlink or single-carrier frequency-domain-equalization (SC-FDE) in the uplink: To exploit the frequency selective channel property without complex equalization.
    - Frequency-domain statistical multiplexing, for example (OFDMA) or (single-carrier FDMA) (SC-FDMA, a.k.a. linearly precoded OFDMA, LP-OFDMA) in the uplink: Variable bit rate by assigning different sub-channels to different users based on the channel conditions.
    - Turbo principle error-correcting codes: To minimize the required SNR at the reception side.
- Channel-dependent scheduling: To use the time-varying channel.
- Link adaptation: Adaptive modulation and error-correcting codes.

- Mobile IP utilized for mobility.

- IP-based femtocells (home nodes connected to fixed Internet broadband infrastructure).

As opposed to earlier generations, 4G systems do not support circuit switched telephony. IEEE 802.20, UMB and OFDM standards lack soft-handover support, also known as cooperative relaying.

## Multiplexing and Access Schemes

Recently, new access schemes like Orthogonal FDMA (OFDMA), Single Carrier FDMA (SC-FDMA), Interleaved FDMA, and Multi-carrier CDMA (MC-CDMA) are gaining more importance for the next generation systems. These are based on efficient FFT algorithms and frequency domain equalization, resulting in a lower number of multiplications per second. They also make it possible to control the bandwidth and form the spectrum in a flexible way. However, they require advanced dynamic channel allocation and adaptive traffic scheduling.

WiMax is using OFDMA in the downlink and in the uplink. For the LTE (telecommunication), OFDMA is used for the downlink; by contrast, Single-carrier FDMA is used for the uplink since OFDMA contributes more to the PAPR related issues and results in non-linear operation of amplifiers. IFDMA provides less power fluctuation and thus requires energy-inefficient linear amplifiers. Similarly, MC-CDMA is in the proposal for the IEEE 802.20 standard. These access schemes offer the same efficiencies as older technologies like CDMA. Apart from this, scalability and higher data rates can be achieved.

The other important advantage of the above-mentioned access techniques is that they require less complexity for equalization at the receiver. This is an added advantage especially in the MIMO environments since the spatial multiplexing transmission of MIMO systems inherently require high complexity equalization at the receiver.

In addition to improvements in these multiplexing systems, improved modulation techniques are being used. Whereas earlier standards largely used Phase-shift keying, more efficient systems such as 64QAM are being proposed for use with the 3GPP Long Term Evolution standards.

## IPv6 Support

Unlike 3G, which is based on two parallel infrastructures consisting of circuit switched and packet switched network nodes, 4G is based on packet switching only. This requires low-latency data transmission.

As IPv4 addresses are (nearly) exhausted, IPv6 is essential to support the large number of wireless-enabled devices that communicate using IP. By increasing the number of IP addresses available, IPv6 removes the need for network address translation (NAT), a method of sharing a limited number of addresses among a larger group of devices, which has a number of problems and limitations. When using IPv6, some kind of NAT is still required for communication with legacy IPv4 devices that are not also IPv6-connected.

As of June 2009, Verizon has posted Specifications that require any 4G devices on its network to support IPv6.

## Advanced Antenna Systems

The performance of radio communications depends on an antenna system, termed smart or intelligent antenna. Recently, multiple antenna technologies are emerging to achieve the goal of 4G systems such as high rate, high reliability, and long range communications. In the early 1990s, to cater for the growing data rate needs of data communication, many transmission schemes were proposed. One technology, spatial multiplexing, gained importance for its bandwidth conservation and power efficiency. Spatial multiplexing involves deploying multiple antennas at the transmitter and at the receiver. Independent streams can then be transmitted simultaneously from all the antennas. This technology, called MIMO (as a branch of intelligent antenna), multiplies the base data rate by (the smaller of) the number of transmit antennas or the number of receive antennas. Apart from this, the reliability in transmitting high speed data in the fading channel can be improved by using more antennas at the transmitter or at the receiver. This is called transmit or receive diversity. Both transmit/receive diversity and transmit spatial multiplexing are categorized into the space-time coding techniques, which does not necessarily require the channel knowledge at the transmitter. The other category is closed-loop multiple antenna technologies, which require channel knowledge at the transmitter.

## Open-wireless Architecture and Software-defined Radio (SDR)

One of the key technologies for 4G and beyond is called Open Wireless Architecture (OWA), supporting multiple wireless air interfaces in an open architecture platform.

SDR is one form of open wireless architecture (OWA). Since 4G is a collection of wireless standards, the final form of a 4G device will constitute various standards. This can be efficiently realized using SDR technology, which is categorized to the area of the radio convergence.

## Disadvantages

4G introduces a potential inconvenience for those who travel internationally or wish to switch carriers. In order to make and receive 4G voice calls, the subscriber handset must not only have a matching frequency band (and in some cases require unlocking), it must also have the matching enablement settings for the local carrier and/or country. While a phone purchased from a given carrier can be expected to work with that carrier, making 4G voice calls on another carrier's network (including international roaming) may be impossible without a software update specific to the local carrier and the phone model in question, which may or may not be available (although fallback to

3G for voice calling may still be possible if a 3G network is available with a matching frequency band).

# WIMAX - REFERENCE NETWORK MODEL

## Subscriber Station/Mobile Station

A mobile station (MS) comprises all user equipment and software needed for communication with a mobile network.

The term refers to the global system connected to the mobile network, i.e. a mobile phone or mobile computer connected using a mobile broadband adapter. This is the terminology of 2G systems like GSM. In 3G systems, a mobile station (MS) is now referred to as user equipment (UE).

In GSM, a mobile station consists of four main components:

- Mobile termination (MT): Offers common functions such as: radio transmission and handover, speech encoding and decoding, error detection and correction, signalling and access to the SIM. The IMEI code is attached to the MT. It is equivalent to the network termination of an ISDN access.

- Terminal equipment (TE): Is any device connected to the MS offering services to the user. It does not contain any functions specific to GSM.

- Terminal adapter (TA): Provides access to the MT as if it were an ISDN network termination with extended capabilities. Communication between the TE and MT over the TA takes place using AT commands.

- Subscriber identity module (SIM): Is a removable subscriber identification token storing the IMSI, a unique key shared with the mobile network operator and other data.

In a mobile phone, the MT, TA and TE are enclosed in the same case. However, the MT and TE functions are often performed by distinct processors. The application processor serves as a TE, while the baseband processor serves as a MT, communication between both takes place over a bus using AT commands, which serves as a TA.

## Base Station (BS)

Base station (or base radio station) is a "land station in the land mobile service."

The term is used in the context of mobile telephony, wireless computer networking and other wireless communications and in land surveying. In surveying, it is a GPS receiver at a known position, while in wireless communications it is a transceiver connecting a number of other devices to one another and/or to a wider area. In mobile telephony, it provides the connection between mobile phones and the wider telephone network. In a computer network, it is a transceiver acting as a switch for computers in the network, possibly connecting them to a/another local area network and/or the Internet. In traditional wireless communications, it can refer to the hub of a dispatch fleet such as a taxi or delivery fleet, the base of a TETRA network as used by government and emergency services or a CB shack.

A 1980s consumer-grade citizens' band radio (CB) base station.

## Land Surveying

In the context of external land surveying, a base station is a GPS receiver at an accurately-known fixed location which is used to derive correction information for nearby portable GPS receivers. This correction data allows propagation and other effects to be corrected out of the position data obtained by the mobile stations, which gives greatly increased location precision and accuracy over the results obtained by uncorrected GPS receivers.

## Computer Networking

In the area of wireless computer networking, a base station is a radio receiver/transmitter that serves as the hub of the local wireless network, and may also be the gateway between a wired network and the wireless network. It typically consists of a low-power transmitter and wireless router.

## Wireless Communications

In radio communications, a base station is a wireless communications station installed at a fixed location and used to communicate as part of one of the following:

- A push-to-talk two-way radio system.

- A wireless telephone system such as cellular CDMA or GSM cell site.

- Terrestrial Trunked Radio.

Base stations use RF power amplifiers (radio-frequency power amplifiers) to transmit and receive signals. The most common RF power amplifiers are metal–oxide–semiconductor field-effect transistors (MOSFETs), particularly LDMOS (power MOSFET) RF amplifiers. RF LDMOS amplifiers replaced RF bipolar transistor amplifiers in most base stations during the 1990s, leading to the wireless revolution.

## Two-way Radio

## Professional

In professional two-way radio systems, a base station is used to maintain contact with a dispatch fleet of hand-held or mobile radios, and/or to activate one-way paging receivers. The base station is one end of a communications link. The other end is a movable vehicle-mounted radio or walkie-talkie. Examples of base station uses in two-way radio include the dispatch of tow trucks and taxicabs.

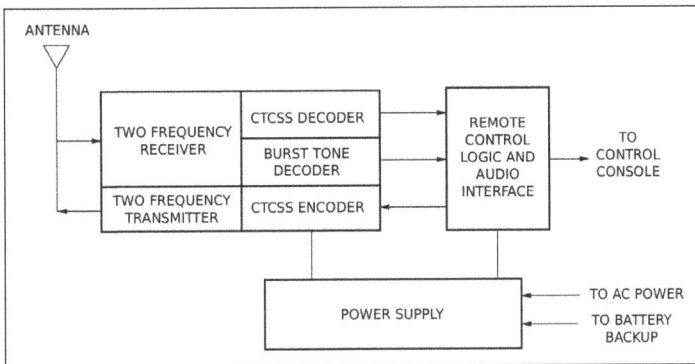

Basic base station elements used in a remote-controlled installation.
Selective calling options such as CTCSS are optional.

Professional base station radios are often one channel. In lightly used base stations, a multi-channel unit may be employed. In heavily used systems, the capability for additional channels, where needed, is accomplished by installing an additional base station for each channel. Each base station appears as a single channel on the dispatch center control console. In a properly designed dispatch center with several staff members, this allows each dispatcher to communicate simultaneously, independently of one another, on a different channel as necessary. For example, a taxi company dispatch center may have one base station on a high-rise building in Boston and another on a different channel in Providence. Each taxi dispatcher could communicate with taxis in either Boston or Providence by selecting the respective base station on his or her console.

In dispatching centers it is common for eight or more radio base stations to be connected to a single dispatching console. Dispatching personnel can tell which channel a message is being received on by a combination of local protocol, unit identifiers, volume settings, and busy indicator lights. A typical console has two speakers identified as select and unselect. Audio from a primary selected channel is routed to the select speaker and to a headset. Each channel has a busy light which flashes when someone talks on the associated channel.

Base stations can be local controlled or remote controlled. Local controlled base stations are operated by front panel controls on the base station cabinet. Remote control base stations can be operated over tone- or DC-remote circuits. The dispatch point console and remote base station are connected by leased private line telephone circuits, (sometimes called RTO circuits), a DS-1, or radio links. The consoles multiplex transmit commands onto remote control circuits. Some system configurations require duplex, or four wire, audio paths from the base station to the console. Others require only a two-wire or half duplex link.

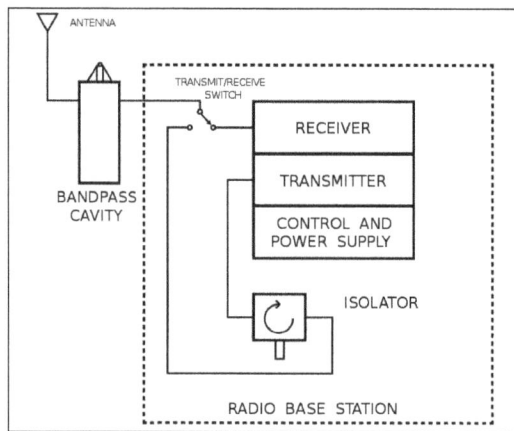

ANTENNA

TRANSMIT/RECEIVE SWITCH

RECEIVER

TRANSMITTER

BANDPASS CAVITY

CONTROL AND POWER SUPPLY

ISOLATOR

RADIO BASE STATION

The diagram shows a band-pass filter used to reduce the base station receiver's exposure to unwanted signals. It also reduces the transmission of undesired signals. The isolator is a one-way device which reduces the ease of signals from nearby transmitters going up the antenna line and into the base station transmitter. This prevents the unwanted mixing of signals inside the base station transmitter which can generate interference.

Interference could be defined as receiving any signal other than from a radio in your own system. To avoid interference from users on the same channel, or interference from nearby strong signals on another channel, professional base stations use a combination of:

*   Minimum receiver specifications and filtering.

*   Analysis of other frequencies in use nearby.

*   In the US, coordination of shared frequencies by coordinating agencies.

*   Locating equipment so that terrain blocks interfering signals.

*   Use of directional antennas to reduce unwanted signals.

Base stations are sometimes called control or fixed stations in US Federal Communications Commission licensing. These terms are defined in regulations inside Part 90 of the commissions regulations. In US licensing jargon, types of base stations include:

- A fixed station is a base station used in a system intended only to communicate with other base stations. A fixed station can also be radio link used to operate a distant base station by remote control. (No mobile or hand-held radios are involved in the system.)

- A control station is a base station used in a system with a repeater where the base station is used to communicate through the repeater.

- A temporary base is a base station used in one location for less than a year.

- A repeater is a type of base station that extends the range of hand-held and mobile radios.

## Amateur and Hobbyist Use

In amateur radio, a base station also communicates with mobile rigs but for hobby or family communications. Amateur systems sometimes serve as dispatch radio systems during disasters, search and rescue mobilizations, or other emergencies.

An Australian UHF CB base station is another example of part of a system used for hobby or family communications.

## Wireless Telephone

Wireless telephone differ from two-way radios in that:

- Wireless telephones are circuit switched: the communications paths are set up by dialing at the start of a call and the path remains in place until one of the callers hangs up.

- Wireless telephones communicate with other telephones usually over the public switched telephone network.

- A wireless telephone base station communicates with a mobile or hand-held phone. For example, in a wireless telephone system, the signals from one or more mobile telephones in an area are received at a nearby base station, which then connects the call to the land-line network. Other equipment is involved depending on the system architecture. Mobile telephone provider networks, such as European GSM networks, may involve carrier, microwave radio, and switching facilities to connect the call. In the case of a portable phone such as a US cordless phone, the connection is directly connected to a wired land line.

## Emissions Issues

A cell tower.

While low levels of radio-frequency power are usually considered to have negligible effects on health, national and local regulations restrict the design of base stations to limit exposure to electromagnetic fields. Technical measures to limit exposure include restricting the radio frequency power emitted by the station, elevating the antenna above ground level, changes to the antenna pattern, and barriers to foot or road traffic. For typical base stations, significant electromagnetic energy is only emitted at the antenna, not along the length of the antenna tower.

Because mobile phones and their base stations are two-way radios, they produce radio-frequency (RF) radiation in order to communicate, exposing people near them to RF radiation giving concerns about mobile phone radiation and health. Hand-held mobile telephones are relatively low power so the RF radiation exposures from them are generally low.

The World Health Organization has concluded that "there is no convincing scientific evidence that the weak RF signals from base stations and wireless networks cause adverse health effects."

The consensus of the scientific community is that the power from these mobile phone base station antennas is too low to produce health hazards as long as people are kept away from direct access to the antennas. However, current international exposure guidelines (ICNIRP) are based largely on the thermal effects of base station emissions, NOT considering the non-thermal effects harmless.

## Emergency Power

Fuel cell backup power systems are added to critical base stations or cell sites to provide emergency power.

Close-up of a base station antenna in Mexico City, Mexico. There are three antennas: each serves a 120-degree segment of the horizon. The microwave dish links the site with the telephone network.

## Base Station Subsystem

The base station subsystem (BSS) is the section of a traditional cellular telephone networkwhich is responsible for handling traffic and signaling between a mobile phone and the network switching subsystem. The BSS carries out transcoding of speech channels, allocation of radio channels to mobile phones, paging, transmission and reception over the air interfaceand many other tasks related to the radio network.

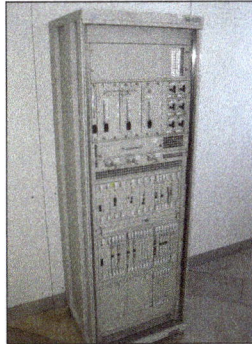

The hardware of GSM base station displayed in Deutsches Museum.

## Base Transceiver Station

Two GSM base station antennas disguised as trees in Dublin, Ireland.

A solar-powered GSM base station on top of a mountain in the wilderness of Lapland.

The base transceiver station, or BTS, contains the equipment for transmitting and receiving radio signals (transceivers), antennas, and equipment for encrypting and decrypting communications with the base station controller (BSC). Typically a BTS for anything other than a picocell will have several transceivers (TRXs) which allow it to serve several different frequencies and different sectors of the cell (in the case of sectorised base stations).

A BTS is controlled by a parent BSC via the "base station control function" (BCF). The BCF is implemented as a discrete unit or even incorporated in a TRX in compact base stations. The BCF provides an operations and maintenance (O&M) connection to the network management system (NMS), and manages operational states of each TRX, as well as software handling and alarm collection.

The functions of a BTS vary depending on the cellular technology used and the cellular telephone provider. There are vendors in which the BTS is a plain transceiver which receives information from the MS (mobile station) through the Um air interface and then converts it to a TDM (PCM) based interface, the Abis interface, and sends it towards the BSC. There are vendors which build their BTSs so the information is preprocessed, target cell lists are generated and even intracell handover (HO) can be fully handled. The advantage in this case is less load on the expensive Abis interface.

The BTSs are equipped with radios that are able to modulate layer 1 of interface Um; for GSM 2G+ the modulation type is Gaussian minimum-shift keying (GMSK), while for EDGE-enabled networks it is GMSK and 8-PSK. This modulation is a kind of continuous-phase frequency shift keying. In GMSK, the signal to be modulated onto the carrier is first smoothed with a Gaussian low-pass filter prior to being fed to a frequency modulator, which greatly reduces the interference to neighboring channels (adjacent-channel interference).

Antenna combiners are implemented to use the same antenna for several TRXs (carriers), the more TRXs are combined the greater the combiner loss will be. Up to 8:1 combiners are found in micro and pico cells only.

Frequency hopping is often used to increase overall BTS performance; this involves the rapid switching of voice traffic between TRXs in a sector. A hopping sequence is followed by the TRXs and handsets using the sector. Several hopping sequences are available, and the sequence in use for a particular cell is continually broadcast by that cell so that it is known to the handsets.

A TRX transmits and receives according to the GSM standards, which specify eight TDMA timeslots per radio frequency. A TRX may lose some of this capacity as some information is required to be broadcast to handsets in the area that the BTS serves. This information allows the handsets to identify the network and gain access to it. This signalling makes use of a channel known as the Broadcast Control Channel (BCCH).

## Sectorization

By using directional antennas on a base station, each pointing in different directions, it is possible to sectorise the base station so that several different cells are served from the same location. Typically these directional antennas have a beamwidth of 65 to 85 degrees. This increases the traffic capacity of the base station (each frequency can carry eight voice channels) whilst not greatly increasing the interference caused to neighboring cells (in any given direction, only a small number of frequencies are being broadcast). Typically two antennas are used per sector, at spacing of ten or more wavelengths apart. This allows the operator to overcome the effects of fading due to physical phenomena such as multipath reception. Some amplification of the received signal as it leaves the antenna is often used to preserve the balance between uplink and downlink signal.

## Base Station Controller

The base station controller (BSC) provides, classically, the intelligence behind the BTSs. Typically a BSC has tens or even hundreds of BTSs under its control. The BSC handles allocation of radio channels, receives measurements from the mobile phones, and controls handovers from BTS to BTS (except in the case of an inter-BSC handover in which case control is in part the responsibility of the anchor MSC). A key function of the BSC is to act as a concentrator where many different low capacity connections to BTSs (with relatively low utilisation) become reduced to a smaller number of connections towards the mobile switching center (MSC) (with a high level of utilisation). Overall, this means that networks are often structured to have many BSCs distributed into regions near their BTSs which are then connected to large centralised MSC sites.

The BSC is undoubtedly the most robust element in the BSS as it is not only a BTS controller but, for some vendors, a full switching center, as well as an SS7 node with connections to the MSC and serving GPRS support node (SGSN) (when using GPRS). It also provides all the required data to the operation support subsystem (OSS) as well as to the performance measuring centers.

A BSC is often based on a distributed computing architecture, with redundancy applied to critical functional units to ensure availability in the event of fault conditions. Redundancy often extends beyond the BSC equipment itself and is commonly used in the power supplies and in the transmission equipment providing the A-ter interface to PCU.

The databases for all the sites, including information such as carrier frequencies, frequency hopping lists, power reduction levels, receiving levels for cell border calculation, are stored in the BSC. This data is obtained directly from radio planning engineering which involves modelling of the signal propagation as well as traffic projections.

## Transcoder

The transcoder is responsible for transcoding the voice channel coding between the coding used in the mobile network, and the coding used by the world's terrestrial circuit-switched network, the Public Switched Telephone Network. Specifically, GSM uses a regular pulse excited-long term prediction (RPE-LTP) coder for voice data between the mobile device and the BSS, but pulse code modulation (A-law or μ-law standardized in ITU G.711) upstream of the BSS. RPE-LPC coding results in a data rate for voice of 13 kbit/s where standard PCM coding results in 64 kbit/s. Because of this change in data rate for the same voice call, the transcoder also has a buffering function so that PCM 8-bit words can be recoded to construct GSM 20 ms traffic blocks.

Although transcoding (compressing/decompressing) functionality is defined as a base station function by the relevant standards, there are several vendors which have implemented the solution outside of the BSC. Some vendors have implemented it in a stand-alone rack using a proprietary interface. In Siemens' and Nokia's architecture, the transcoder is an identifiable separate sub-system which will normally be co-located with the MSC. In some of Ericsson's systems it is integrated to the MSC rather than the BSC. The reason for these designs is that if the compression of voice channels is done at the site of the MSC, the number of fixed transmission links between the BSS and MSC can be reduced, decreasing network infrastructure costs.

This subsystem is also referred to as the transcoder and rate adaptation unit (TRAU). Some networks use 32 kbit/s ADPCM on the terrestrial side of the network instead of 64 kbit/s PCM and the TRAU converts accordingly. When the traffic is not voice but data such as fax or email, the TRAU enables its rate adaptation unit function to give compatibility between the BSS and MSC data rates.

## Packet Control Unit

The packet control unit (PCU) is a late addition to the GSM standard. It performs some of the processing tasks of the BSC, but for packet data. The allocation of channels between voice and data is controlled by the base station, but once a channel is allocated to the PCU, the PCU takes full control over that channel.

The PCU can be built into the base station, built into the BSC or even, in some proposed architectures, it can be at the SGSN site. In most of the cases, the PCU is a separate node communicating extensively with the BSC on the radio side and the SGSN on the Gb side.

## BSS Interfaces

## Um

The air interface between the mobile station (MS) and the BTS. This interface uses LAPDm protocol for signaling, to conduct call control, measurement reporting, handover, power control, authentication, authorization, location update and so on. Traffic and signaling are sent in bursts of 0.577 ms at intervals of 4.615 ms, to form data blocks each 20 ms.

Image of the GSM network, showing the BSS interfaces to the MS, NSS and GPRS Core Network.

## Abis

The interface between the BTS and BSC. Generally carried by a DS-1, ES-1, or E1 TDM circuit. Uses TDM subchannels for traffic (TCH), LAPD protocol for BTS supervision and telecom signaling, and carries synchronization from the BSC to the BTS and MS.

## A

The interface between the BSC and MSC. It is used for carrying traffic channels and the BSSAP user part of the SS7 stack. Although there are usually transcoding units between BSC and MSC, the signaling communication takes place between these two ending points and the transcoder unit doesn't touch the SS7 information, only the voice or CS data are transcoded or rate adapted.

## Ater

The interface between the BSC and transcoder. It is a proprietary interface whose name depends on the vendor (for example Ater by Nokia), it carries the A interface information from the BSC leaving it untouched.

## Gb

Connects the BSS to the SGSN in the GPRS core network.

## Access Service Network Gateway (ASN-GW)

Access service networks (ASN) provide a means to connect mobile subscribers using OFDMA air link to IP backbone with session continuity. ASN comprises base stations (BS) and access gateways named ASN-GW. The interface between the ASN and mobile subscriber is through BS with IEEE 802.16e-2005 standard.

## Access Service Network Functional Protocols

Protocol Layering of WiMAX considers end-to-end protocol layering. Data and control packets are forwarded from the MS to the CSN in uplink. The traffic is concentrated in the ASN-GW and forwarded to the CSN and same way, concentrated in the ASN-GW for downlink and distributed to the MSs residing in different BSs. IP packets use IP convergence sublayer (IP-CS) or Ethernet convergence sublayer (ETH-CS) over IEEE 802.16e. The IP-CS with IP-in-IP encapsulation between BS and ASN-GW is considered in most designs. Bridging is also another way of routing packet within ASN.

Network Discovery and Selection implements manual or automatic selection of the appropriate network. MS first discovers all the NAPs where each has an Operator ID embedded into Base Station ID and transmitted with DL-MAP of each frame. And MS continue to listen the channel for SII-ADV signal which system identity information advertisement to advertise NSP IDs. The MS selects one NSP from the list according to an algorithm and performs network entry and provide its identity and its home NSP domain with a network access identifier (NAI). The ASN selects the next AAA hop from the realm portion of the NAI.

IP Address Assignment is done through DHCP or AAA: ASN hosts DHCP relay or DHCP proxy respectively. In order to deliver the point of attachment IP address to MS. For IPv6 there is access router in ASN to obtain globally routable IP address. The MS gets the care-of-address (CoA) from ASN and home address (HoA) from CSN.

Authentication and Security Architecture implements 802.16e security with IETF EAP framework. AAA framework is used for service flow authorization, mobility management and policy control. AAA framework is based on pull model in which supplicant sends a request to ASN and ASN forwards it to AAA server. The AAA return with appropriate response to ASN which set up the service and inform the MS.

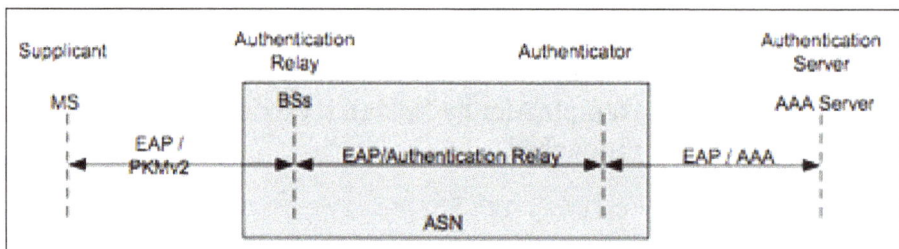

Authentication relay inside the ASN.

User and device authentication is supported with PKMv2 and EAP. PKMv2 is between MS and BS and BS relays this EAP messages to ASN-GW where AAA client encapsulates the EAP and forwards to AAA server in the CSN over RADIUS. EAP-AKA, EAP-TLS, EAP-SIM, EAP-PSK, EAP-TTLS are the supported EAP types. Both user and device authentication is performed with double-EAP and device credentials are in the form of digital certificate, secret key, or X.509 certificate.

Quality of Service Architecture in WiMAX complements the QoS framework in IEEE 802.16e-2005 QoS model. The QoS provides rich set of variety: per user and per service flow basis differentiated levels; admission control; bandwidth optimization. QoS provides static and dynamic service flow creation. For each service there is provisioned, admitted, and active states. When flow is in active state it starts getting the service. Entities are Policy function and AAA server residing in CSN, Service flow management residing in BS, and Service flow authorization residing in ASN-GW.

Mobility Management implements mobility with the ASN and across the ASNs. ASN-anchored mobility is when MS moves within the same Foreign Agent domain residing in ASN-GW. Control signals use R6 and R8 reference points and data path shift happens in ASN-GW with new R6 to target BS when handover is complete. CSN-anchored mobility additional to ASN-anchored mobility triggers the FA change through Home Agent. Now, R3 and R4 reference points also become active.

Radio Resource Management is responsible to fully utilize the network by information gathering and implementing decisions. The information such as radio-related measurements; base station spare capacity reports are concentrated to assist handover decision and load balancing decisions.

Paging and Idle Mode Operation is responsible to maintain a track and alert for MS when it is in idle mode for battery power saving reasons. Paging is executed to alert MS when there is an incoming message. MS is tracked when it is in the idle mode and information is stored to a location register (LR). Granularity of track is bigger than cell size since a paging group (PG) is composed of multiple cell and when a MS moves across paging groups, location update occurs via R6 and/or R4. Paging Controller (PG) in ASN-GW retrieves the location from LR and alerts the paging agent in (PA) in BS to signal to MS.

# Protocol Layers and Topologies

WiMAX physical layers are of five types namely SC, SCa, HUMAN, OFDM and OFDMA. A point-to-multipoint topology is operated using LOS signal propagation for providing network access from one location to another. The topics elaborated in this chapter will help in gaining a better perspective about protocol layers and topologies.

## WIMAX PROTOCOL

WiMax has two main topologies namely Point to Point for backhaul and Point to Multi Point Base station for Subscriber station.

In each of these situations, multiple input multiple output antennas are used. The four layers Convergence, MAC, Transmission and Physical. These layers map to two of the lowest layers physical and data link layers of the OSI model.

WiMax provides many user applications and interfaces like Ethernet, TDM, ATM, IP, and VLAN.

The IEEE 802.16 standard is versatile enough to accommodate time division multiplexing (TDM) or frequency division duplexing (FDD) deployments and also allows for both full an d half-duplex terminals.

802.16 supports three physical layers. The mandatory physical mode is 256-point FFT OFDM (Orthogonal Frequency Division Multiplexing). The other modes are Single carrier (SC) and 2048 OFDMA (Orthogonal Frequency Division Multiplexing Access) modes. The corresponding European standard - the ETSI Hiperman standard defines a single PHY mode identical to the 256 OFDM modes in the 802.16d standard.

The MAC was developed for a point-to-multipoint wireless access environment and can accommodate protocols like ATM, Ethernet and IP (Internet Protocol). The MAC frame structure dynamic uplink and downlink profiles of terminals as per the link conditions. This is to ensure a trade-off of capacity and real-time robustness.

The MAC uses a protocol data unit of variable length, which increases the standards efficiency. Multiple MAC protocol data unit can be sent as a single physical stream to save overload. Also, multiple Service data units (SDU) can be sent together to save on

MAC header overhead. By fragmenting, you can send large volumes of data (SDUs) across frame boundaries and can guarantee a QoS (Quality of Service) of competing services. The MAC uses a self-correcting bandwidth request scheme to avoid overhead and acknowledgement delays.

This also allows better QoS handling than the traditional acknowledged schemes. The terminals have a variety of options to request for bandwidth depending on the QoS and other parameters. The signal requirement can be polled or a request can be piggy-backed.

The 802.16 MAC protocol performs mainly two tasks Periodic and Aperiodic activities. Fast activities (periodic) like scheduling, packing, fragmentation and ARQ are hard-pressed for time and have hard tight deadlines. They must be performed within a single frame.

The slow activities, on the other hand, typically execute as per pre-fixed timers, but are not associated with any timers. They also do not have specific time frame or deadline.

# PROTOCOL LAYERS OF WIMAX

The WiMAX MAC layer, or IEEE 802.16 MAC is an essential elements within the overall WiMAX software stack. These elements enable WiMAX to perform as an effective wireless broadband system.

The WiMAX MAC layer is a form of MAC used for the WiMAX system.

A MAC layer or Media Access Control data communication protocol sub-layer may also be known as a Medium Access Control layer. WiMax Protocol

A MAC layer is a sub-layer of the Data Link Layer. This is defined in the standard seven-layer OSI model as layer 2. The MAC layer provides addressing and channel access control mechanisms that make it possible for several terminals or network nodes to communicate within a multi-point network, typically a local area network (LAN) or metropolitan area network (MAN).

The WiMAX MAC has been designed and optimised to enable point to multipoint wireless applications and the WiMAX MAC layer provides an interface between the physical layer and the higher application layers within the stack.

The WiMAX MAC layer has to meet a number of requirements:

- Point to multipoint: One of the main requirements for WiMAX is that it must be possible for a base station to communicate with a number of different outlying users, either fixed or mobile. To achieve this, the IEEE 802.16, WiMAX

MAC layer is based on collision sense multiple access with collision avoidance, CSMA/CA to provide the point to multipoint, PMP capability.

- Connection orientated.

- Supports communication in all conditions: The WiMAX MAC layer must be able to support a large number of users along with high data rates. As the traffic is packet data orientated it must be able to support both continuous and" bursty" traffic. Most data traffic is "bursty" in nature having short times of high data rates then remaining dormant for a short while.

- Efficient spectrum use: The WiMAX MAC must be capable of supporting methods that enable very efficient use of the spectrum.

- Variety of QoS options: To provide the support for different forms of traffic from voice data to Internet surfing, etc, a variety of different classes and forms of QoS support are needed. Support for QoS is a fundamental part of the WiMAX MAC-layer. The WiMAX MAC utilises some of the concepts that are embedded in the DOCSIS cable modem standard.

- Multiple WiMAX/IEEE 802.16 physical layers: With different variants, the WiMAX MAC layer must be able to provide support for the different PHYs.

## WiMAX MAC Layer Operation

The WiMAX MAC layer is primarily an adaptation layer between the physical layer and the upper layers within the overall stack.

One of the main tasks of the WiMAX MAC layer is to transfer data between the various layers.

- Transmission of data: Reception of MAC Service Data Units, MSDUs from the layer above. It then aggregates and encapsulates them into MAC Protocol Data Units, MPDUs, before passing them to the physical layer, PHY for transmission.

- Reception of data: The WiMAX MAC layer takes MPDUs from the physical layer. It decapsulates and reorganises them into MSDUs, and then passes them on to the upper-layer protocols.

For the different formats: IEEE 802.16-2004 and IEEE 802.16e-2005, the WiMAX MAC design includes a convergence sublayer. This is used to interface with a variety of higher-layer protocols, such as ATM, Ethernet, IP, TDM Voice, and other future protocols that may arise.

WiMAX defines a concept of a service flow and has an accompanying Service Flow Identifier, SFID. The service flow is a unidirectional flow of packets with a particular set of QoS parameters, and the identifier is used to identify the flow to enable operation.

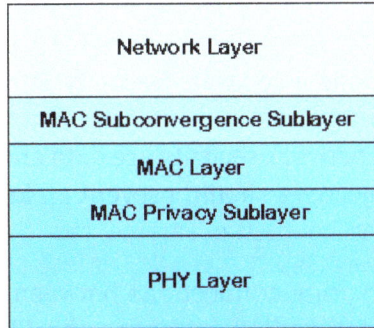

WiMAX Protocol Stack.

There is an additional layer between the WiMAX MAC itself and the upper layers. This is called the Convergence Sublayer. For the upper protocol layers, the convergence sublayer acts as an interface to the WiMAX MAC. Currently the convergence sublayer only supports IP and Ethernet, although other protocols can be supported by encapsulating the data.

The WiMAX MAC layer provides for a flexible allocation of capacity to different users. It is possible to use variably sized MPDUs from different flows - these can be included into one data burst before being handed over to the PHY layer for transmission. Also, multiple small MSDUs can be aggregated into one larger MPDU. Conversely, one big MSDU can be fragmented into multiple small ones in order to further enhance system performance. This level of flexibility gives significant improvements in overall efficiency.

## WiMAX MAC Connection Identifier

Before any data is transferred over a WiMAX link, the user equipment or mobile station and the base station must create a connection between the WiMAX MAC layers of the two stations. To achieve this, an identifier known as a Connection Identifier, CID, is generated and assigned to each uplink/downlink connection. The CID serves as an intermediate address for the data packets transmitted over the WiMAX link.

There is another identifier used within the WiMAX MAC layer. Known as the Service Flow Identifier, SFID, this is assigned to unidirectional packet data traffic by the base station. It is worth noting that the base station WiMAX MAC layer also handles the mapping of the SFIDs to CIDs to provide the required quality of service.

The WiMAX MAC layer also incorporates a number of other features including power-management techniques and security features.

The WiMAX MAC layer has been developed to provide the functionality required for a point to multipoint system and to provide wireless broadband. The WiMAX MAC layer is also able to provide support for the different physical layers needed for the different flavours of WiMAX that are in use.

## Security Sublayer

The MAC Sublayer also contains a separate Security Sublayer providing authentication, secure key exchange, encryption and integrity control across the BWA system. The two main topics of a data network security are data encryption and authentication. Algorithms realising these objectives should prevent all known security attacks whose objectives may be denial of service, theft of service, etc.

In the 802.16 standard, encrypting connections between the SS and the BS is made with a data encryption protocol applied for both ways. This protocol defines a set of supported cryptographic suites, i.e. pairings of data encryption and authentication algorithms. An encapsulation protocol is used for encrypting data packets across the BWA. This protocol defines a set of supported cryptographic suites, i.e. pairings of data encryption and authentication algorithms. The rules for applying those algorithms to an MAC PDU payload are also given.

An authentication protocol, the Privacy Key Management (PKM) protocol is used to provide the secure distribution of keying data from the BS to the SS. Through this secure key exchange, due to the key management protocol the SS and the BS synchronize keying data. The basic privacy mechanisms are strengthened by adding digital-certificate-based SS authentication to the key management protocol. In addition, the BS uses the PKM protocol to guarantee conditional access to network services. The 802.16e amendment defined PKMv2 which has the same framework as PKM, re-entitled PKMvl, with some additions such as new encryption algorithms, mutual authentication between the SS and the BS, support for a handover and a new integrity control algorithm.

## Security Elements used in the 802.16 Standard

A wireless system uses the radio channel, which is an open channel. Hence, security procedures must be included in order to protect the traffic confidentiality and integrity and to prevent different network security attacks such as theft of service. The IEEE 802.16 MAC layer contains a security sublayer.

The Privacy Key Management (PKM) protocol, later known as PKMvl, is included in the 802.16-2004 standard security sublayer in order to provide secure distribution of keying data from the BS to the SS. In addition, PKM is used to apply conditional access to network services, making it the authentication protocol, protecting them from theft of service (or service hijacking) and providing a secure key exchange. Many ciphering (data encryption) algorithms are included in the 802.16 standard security sublayer for encrypting packet data across the 802.16 network.

The Security sublayer has been redefined in the IEEE 802.16e amendment mainly due to the fact that 802.16-2004 had some security holes (e.g. no authentication of the BS) and that the security requirements for mobile services are not the same as for fixed services. The Security sublayer has two main component protocols as follows:

- A data encapsulation protocol for securing packet data across the fixed BWA network. This protocol defines a set of supported cryptographic suites, i.e. pairings of data encryption and authentication algorithms, and the rules for applying those algorithms to a MAC PDU payload.

- A key management protocol (PKM) providing the secure distribution of keying data from the BS to the SS. Through this key management protocol, the SSs and BSs synchronise keying data. In addition, the BS uses the protocol to enforce conditional access to network services. The 802.16e amendment defined PKMv2 with enhanced features.

## Encryption Algorithms

Many encryption algorithms are included in the 802.16 standard Security sublayer. They can be used for securing ciphering key exchange and for the encryption of transport data. Some of these algorithms are optional for some applications.

The encryption algorithms included in 802.16 are:

- RSA (Rivest Shamir Adleman): RSA is a public-key asymmetric encryption algorithm used to encrypt the Authorisation Reply message using the SS public key. The Authorisation Reply message includes the Authorisation Key (AK). RSA may also be used for the encryption of traffic encryption keys when these are transmitted from the BS to the SS.

- DES (Data Encryption Standard): The DES and 3-DES are shared(secret)-key encryption algorithms. The DES algorithm may be used for traffic data encryption. It is mandatory for 802.16 equipment. The 3-DES algorithm can be used for the encryption of the traffic encryption keys.

- AES (Advanced Encryption Standard): The AES algorithm is a shared(secret)-key encryption algorithm. The AES algorithm may be used for traffic data encryption and can also be used for the encryption of the traffic encryption keys. Its implementation is optional.

Cryptographic algorithms are also included in 802.16.

- HMAC (Hashed Message Authentication Code)and CMAC (Cipher-based Message Authentication Code): HMAC and CMAC are used for message authentication and integrity control.

## X.509 Certificate

ITU-T X.509 (formerly CCITT X.509) or ISO/IEC 9594-8, which was first published in

1988 as part of the X.500 directory recommendations, defines a standard certificate format used in IETF RFC 3280, itself used in the 802.16 standard (citing RFC 2459 of the IETF).

The 802.16 standard states that 802.16-compliant SSs must use X.509 Version 3 certificate formats providing a public key infrastructure used for secure authentication. Each SS carries a unique X.509 digital certificate issued by the SS manufacturer, known as the SS X.509 certificate. More exactly, this certificate is issued (and signed) by a Certification Authority (CA) and installed by the manufacturer. This digital certificate contains the SS RSA public key and the SS MAC address.

Each SS has a manufacturer-issued X.509 manufacturer CA certificate issued by the manufacturer or by an external authority. The manufacturer's public key is then placed in this X.509 manufacturer CA certificate, which in turn is signed by a higher-level CA. This higher-level CA does not seem to be clearly defined in the present version of the standard. There are then two types of X.509 certificates: SS X.509 certificates and the X.509 manufacturer CA certificate. In the 802.16-2004 standard, there is an X.509 certificate for the BS.

The main fields of the X.509 certificate are the following:

- The X.509 certificate version is always set to v3 when used in the 802.16 standard.

- The certificate serial number is a unique integer the issuing CA assigns to the certificate.

- The signature algorithm is an object identifier and optional parameters defining the algorithm used to sign the certificate.

- The signature value is the digital signature computed on the ASN.1 DER encoded tbsCertificate ('to be signed Certificate'). The standard states that the RSA signature algorithm with SHA-1 (Secure Hash Algorithm), is employed for both defined certificate types.

- The certificate issuer is the Distinguished Name (DN) of the CA that issued the certificate.

- The certificate validity is when the certificate becomes active and when it expires.

- The certificate subject is the DN identifying the entity whose public key is certified in the subject public key information field. The country name, organisation name and manufacturing location are attributes of the certificate subject. Another attribute is the company name for the CA or, for the SS, its 48-bit universal MAC address and its serial number. This MAC address is used to identify the SS to the various provisioning servers during initialisation.

- The Certificate Subject Public Key Info Field contains the public key material (public key and parameters) and the identifier of the algorithm with which the key is used. This is the main content of the X.509 certificate.

- Optional fields allow reuse of issuer names over time.

- The certificate extensions are extension data.

The BS uses the X.509 certificate public key of an SS in order to verify the authenticity of this certificate and then authenticates the SS. This is done using the PKM protocol. This security information also authenticates the responses received by the SS.

## Encryption Keys and Security Associations (SAs)

The standard defines a Security Association (SA) as a set of security information a BS and one or more of its client SSs (or MSs) share in order to support secure communications. An SA's shared information includes the Cryptographic Suite employed within the SA. A Cryptographic Suite is the SA's set of methods for data encryption, data authentication and TEK exchange. The exact content of the SA is dependent on the SA's Cryptographic Suite: encyption keys, keys lifetime, etc. The Security Association Identifier (SAID) is a 16-bit identifier shared between the BS and the SS that uniquely identifies an SA. There are three types of security associations: the SA for unicast connections, the Group Security Association (GSA) for multicast groups and the MBS Group Security Association (MBSGSA) for MBS services.

The SAs are managed by the BS. When an authentication event takes place the BS gives the SS a list of Security Associations associated with its connections. Generally, an SS has a Security Association ('primary' association) for its secondary management connection and two more for the downlink and the uplink links. After that, the BS may indicate one or more new SAs to the SS.

# WIMAX TOPOLOGIES

## PMP (Point-to-Multipoint) Topology

The Point-to-Multipoint topology (also called star topology or simply P2MP) is a common network architecture for outdoor wireless networks to connect multiple locations to one single central location. In a point-to-multipoint wireless Ethernet network, all remote locations do not communicate directly with each other but have a single connection towards the center of the star network where one or more base station is typically located.

The remote locations at the edge of the networks are typically called "client" locations and the central location is called the "access point" or "base station". Point-to-multipoint wireless networks have been studied in the 1990s and in the early 2000s and discussed in many academic publications because they can be affected by certain issues such as the hidden terminal issues or the exposed terminal issues, depending on the point-to-multipoint protocol implemented to coordinate the transmissions over the wireless medium. Most outdoor point-to-multipoint networks implement a centralised medium access control protocol or employ a TDMA-based protocol synchronising all radio devices with a GPS device in order to avoid the hidden terminal and exposed terminal issues.

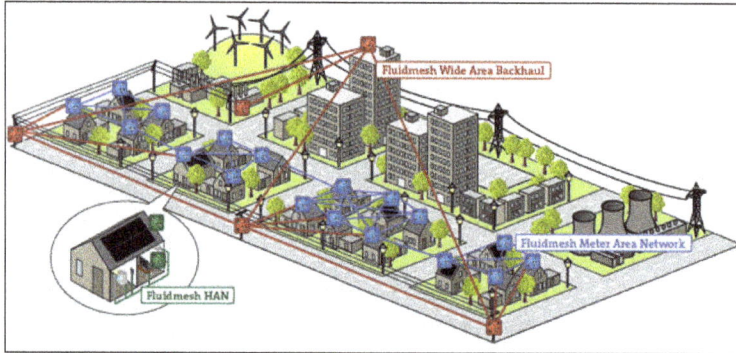

Point-to-Multipoint Wireless Network.

Outdoor point-to-multipoint wireless solutions are very common both for wireless Internet service providers (WISPs) and for outdoor video-surveillance systems. In a WISP network, subscribers are connected at the edge of the network using a client device typically mounted on the roof of their house. One or more central base stations are mounted on a high building, or on a mountaintop or on a water tower in line of sight with as many client devices as possible.

In outdoor wireless video-surveillance systems, each camera in the field is connected to a wireless client device and then a base station is mounted on top of a tall building and acts as the central device and coordinator of the point-to-multipoint wireless network. In a point-to- multipoint wireless CCTV system, all video streams from the remote cameras are collected at this central location at the center of the point-to-multipoint wireless system and then transmitted to a control room using a point-to-point wireless or fiber backhaul.

Point-to-multipoint wireless links are deployed between locations where the client wireless devices  are in clear line of sight (LOS) with the device acting as the base station. In a point-to-multipoint wireless network that works using the license free 5 GHz band (for example in the 5.8 GHz or 5.4 GHz license-free bands) or using the 4.9 GHz public safety band we suggest to deploy the link in clear ling of sight (LOS) because, above  2.4 GHz, line of sight (LOS) operations provide more reliable performances. Point to multipoint networks working on frequencies around 900 MHz or in the UHF band (400 MHz) can operate reliably in near line of sight (NLOS) or in non line of sight conditions (NLOS point to point wireless links).

Fluidmesh Networks provides advanced outdoor point-to-multipoint wireless solutions for long range and high throughput applications.

The Fluidmesh 3200 Base is the core of the point-to-multipoint Ethernet system. The radio has a 120 degree sector point-to-multipoint antenna and can act as the central base station and access point in point-to-multipoint wireless networks. The Fluidmesh 3200 Base is certified to work in the 5 GHz band (including in the 5.4 GHz and 5.8 GHz frequency band) and in the 4.9 GHz public safety band.

The Fluidmesh 1200 VOLO can be deployed at the edge of the point-to-multipoint wireless Ethernet network and used to connect remote IP cameras or other Ethernet devices such as sensors, VOIP phone or entire IP/Ethernet networks. The Fluidmesh 1200 Volo is certified to work in the 5 GHz band (including in the 5.4 GHz and 5.8 GHz frequency band) and in the 4.9 GHz public safety band.

## References

- Protocol, wireless-wimax: tutorial-reports.com, Retrieved 16 July, 2019

- Mac-layer, connectivity-wimax: electronics-notes.com, Retrieved 23 May, 2019

- Security-Sublayer, wimax-technology-broadband-wireless-access-Part-One-Global-Introduction-to-WiMAX-Protocol-Layers-and-Topologies: etutorials.org, Retrieved 19 June, 2019

- Security, wimax-technology-broadband-wireless-access: etutorials.org, Retrieved 28 April, 2019

- Point-to-multipoint-wireless: fluidmesh.com, Retrieved 01 February, 2019

# Wireless Display Technologies

Wireless display technologies enable the users to stream music, movies, photos, videos and applications without wires from a compatible computer to a compatible HDTV. Some of these technologies are Chromecast, Google cast, EZCast, Micracast, OpenFlint, etc. This chapter closely examines these wireless display technologies to provide an extensive understanding of the subject.

## WIRELESS DISPLAY

A wireless display is any type of display – i.e. flat panel LED, LCD, projector, video wall, etc. – that can be accessed wirelessly from a separate device – such as a laptop, tablet or smartphone. The vast majority of the wireless display solutions available in the market operate over standard IP networks like WiFi. In other words, users join the WiFi network that the wireless display is attached to in order to connect. In general, today's enterprise wireless display solutions are separate consoles or dongles that plug into existing displays to make the displays wirelessly accessible.

At the most basic level, wireless displays enable users to share content from a device to the display without being tethered to the display by way of a video cable. If you've ever walked into a conference room to give a presentation, you probably had to plug an HDMI or VGA cable into your laptop in order to show your presentation up on the main screen. Wireless displays cut the cable in conference rooms, enabling users to present on the main screen wirelessly.

Beyond cutting the cable, wireless display solutions have fewer things in common than you might guess. Each solution has a unique approach to the problem and supports different features. At the highest level, we should distinguish between consumer solutions – that primarily serve entertainment purposes – and productivity-focused, enterprise wireless display solutions employed by businesses and education institutions.

Consumer solutions are primarily used for streaming entertainment content like Netflix. These solutions are generally limited to one connected user at a time, and often have limited support for the various user device platforms, such as support for Apple AND Windows devices. A couple examples of these solutions include Google

Chromecast and Apple TV. Ultimately these consumer products can be great for home/consumer use but usually aren't the best solutions for meeting rooms or classrooms.

On the other hand, enterprise wireless display solutions are productivity-focused and usually support a broader range of content (like business applications, presentations, etc.) as well as a broader range of user devices (like Windows, Apple, and Android). However even within the 'productivity-focused' category, there is a lot of distinction and variation between solutions in terms of features and the overall approach to wireless displays.

Here are a few factors that are the most important and distinguish our Solstice wireless display solution from the pack.

- Unlimited users with unlimited sharing. The single biggest factor that sets Solstice apart from other wireless display solutions is that Solstice supports any number of connected users sharing any amount of content on the display simultaneously. Any number of Solstice users can connect and share any amount of content at once, supporting any type of meeting – from a single-presenter session, to an auditorium full of collaborators each sharing content simultaneously.

- Customizable layouts and user control. In addition to supporting unlimited users and sharing, only Solstice gives connected users control of both the media content shared (e.g. any users can pause or play a video shared by another user) AND control of the layout of the content on the screen. Users can arrange, move, delete, and scale content posts to achieve the layout that best serves their particular meeting. The result is more engaged meeting participants and higher fidelity results based on user-controlled content and layouts customized for the task at hand.

- Future-Proof Software Architecture. Solstice is a software-based wireless display solution which means that, in addition to its unique features, new features are added quickly and frequently and are accessible via over-the-air software updates. This is really important for an emerging technology like wireless displays because user requirements are still being defined. With Solstice, customers aren't limited to the features available on the date of purchase. Additionally, the software-based wireless display solution leverages previous investments in the meeting room equipment and infrastructure, such as existing in-room PCs and WiFi/Ethernet networks.

## WIDI

Wireless Display (WiDi) was technology developed by Intel that enabled users to stream music, movies, photos, videos and apps without wires from a compatible computer to a compatible HDTV or through the use of an adapter with other HDTVs or monitors. Intel WiDi supported HD 1080p video quality, 5.1 surround sound, and low latency for interacting with applications sent to the TV from a PC.

Using the Intel WiDi Widget users could perform different functions simultaneously on their PC and TV such as checking email on the PC while streaming a movie to the TV from the same device.

WiDi was discontinued in 2015 in favour of Miracast, a standard developed by the Wi-Fi Alliance and natively supported by Windows 8.1 and later.

# WIRELESS HOME DIGITAL INTERFACE

Wireless Home Digital Interface (WHDI) is a consumer electronic specification for a wireless HDTV connectivity throughout the home.

WHDI enables delivery of uncompressed high-definition digital video over a wireless radio channel connecting any video source (computers, mobile phones, Blu-ray players etc.) to any compatible display device. WHDI is supported and driven by Amimon, Hitachi Ltd., LG Electronics, Motorola, Samsung Group, Sharp Corporation and Sony.

## Versions

The WHDI 1.0 specification was finalized in December 2009. Sharp Corporation will be one of the first companies to roll out wireless HDTVs. AT CES 2010 LG Electronics announced a WHDI wireless HDTV product line.

In June 2010, WHDI announced an update to WHDI 1.0 which allows support for stereoscopic 3D, and WHDI 2.0 specification to be completed in Q2 2011.

WHDI 3D update due in Q4 2010 will allow support for 3D formats defined in HDMI 1.4a specification

WHDI 2.0 will increase available bandwidth even further, allowing additional 3D formats such as "dual 1080p60", and support for 4K × 2K resolutions.

## Technology

WHDI 1.0 provides a high-quality, uncompressed wireless link which supports data rates of up to 3 Gbit/s (allowing 1920×1080 @ 60 Hz @ 24-bit) in a 40 MHz channel, and data rates of up to 1.5 Gbit/s (allowing 1280×720 @ 60 Hz @ 24-bit or 1920×1080 @ 30 Hz @ 24-bit) in a single 20 MHz channel of the 5 GHz unlicensed band, conforming to FCC and worldwide 5 GHz spectrum regulations. Range is beyond 100 feet (30 m), through walls, and latency is less than one millisecond.

# WIRELESSHD

WirelessHD, also known as UltraGig, is a proprietary standard owned by Silicon Image (originally SiBeam) for wireless transmission of high-definition video content for consumer electronics products. The consortium currently has over 40 adopters; key members behind the specification include Broadcom, Intel, LG, Panasonic, NEC, Samsung, SiBEAM, Sony, Philips and Toshiba. The founders intend the technology to be used for Consumer Electronic devices, PCs, and portable devices.

## Technology

The WirelessHD specification is based on a 7 GHz channel in the 60 GHz Extremely High Frequency radio band. It allows either lightly compressed (proprietary wireless link-aware codec) or uncompressed digital transmission of high-definition video and audio and data signals, essentially making it equivalent of a wireless HDMI. First-generation implementation achieves data rates from 4 Gbit/s, but the core technology allows theoretical data rates as high as 25 Gbit/s (compared to 10.2 Gbit/s for HDMI 1.3 and 21.6 Gbit/s for DisplayPort 1.2), permitting WirelessHD to scale to higher resolutions, color depth, and range. The 1.1 version of the specification increases the maximum data rate to 28 Gbit/s, supports common 3D formats, 4K resolution, WPAN data, low-power mode for portable devices, and HDCP 2.0 content protection.

The 60 GHz band usually requires line of sight between transmitter and receiver, and the WirelessHD specification ameliorates this limitation through the use of beam forming at the receiver and transmitter antennas to increase the signal's effective radiated power, find the best path, and utilise wall reflections. The goal range for the first products will be in-room, point-to-point, non line-of-sight (NLOS) at up to 10 meters. The atmospheric absorption of 60 GHz energy by oxygen molecules limits undesired propagation over long distances and helps control intersystem interference and long distance reception, which is a concern to video copyright owners.

The WirelessHD specification has provisions for content encryption via Digital Transmission Content Protection (DTCP) as well as provisions for network management. A standard remote control allows users to control the WirelessHD devices and choose which device will act as the source for the display.

## Competition

WirelessHD competes with WiGig in some applications. WiGig transmits in the same 60 GHz band used by WirelessHD.

## AIRPLAY

AirPlay is a proprietary protocol stack/suite developed by Apple Inc. that allows wireless streaming between devices of audio, video, device screens, and photos, together with related metadata. Originally implemented only in Apple's software and devices, it was called AirTunes and used for audio only. Apple has since licensed the AirPlay protocol stack as a third-party software component technology to manufacturers that build products compatible with Apple's devices.

Apple announced AirPlay 2 at its annual WWDC conference on June 5, 2017. It was scheduled for release along with iOS 11 in the third quarter of 2017, but was delayed until June 2018. Compared to the original version, AirPlay 2 improves buffering; adds streaming audio to stereo speakers; allows audio to be sent to multiple devices in different rooms; and control by Control Center, the Home app, or Siri, functionality that was only available previously using iTunes under macOS or Windows.

### Senders

AirPlay sender devices include computers running iTunes, and iOS devices such as iPhones, iPods, and iPads running iOS 4.2 or greater, and devices can send AirPlay over wi-fi or ethernet. OS X Mountain Lion supports display mirroring via AirPlay on systems containing 2nd generation Intel Core processors or later.

In 2016, HTC released an Android phone with Apple AirPlay streaming.

As of iOS 4.3, third-party apps may send compatible audio and video streams over AirPlay. The iTunes Remote app on iOS can be used to control media playback and select AirPlay streaming receivers for iTunes running on a Mac or PC.

As of macOS 10.14, there is no public API for third-party developers to integrate AirPlay 2 into their macOS apps. However, there are third-party streamers such as Airfoil. In May 2019, a third-party developer released a macOS app that can stream audio using AirPlay 2. The app includes a helper tool called "AirPlay Enabler" that uses code injection to bypass restrictions to the AirPlay 2 private API on macOS.

### Receivers

AirPlay receiver devices include Apple TV, HomePod, other third-party speakers and the discontinued AirPort Express, which included a combined analog and optical S/PDIF audio output connector. Compatible devices can receive AirPlay over wi-fi or ethernet. With the open-source implementations of the AirPlay protocol, any computer can be turned into an AirPlay receiver.

However, because not all third-party receivers implement Apple's DRM encryption, some media, such as iTunes Store's own rights-protected music (Apple's own "FairPlay" encryption), YouTube, and Netflix, cannot stream to those devices or software. On Apple TV, starting with firmware 6.0, the DRM scheme is enforced: devices without it cannot be used.

AirPlay wireless technology is integrated into speaker docks, AV receivers, and stereo systems from companies such as Bose, Yamaha, Philips, Marantz, Onkyo, Bowers & Wilkins, Pioneer, Sony, Sonos, McIntosh, Denon, and Bang & Olufsen. Song titles, artists, album names, elapsed and remaining time, and album artwork can appear on Air-Play-enabled speakers with graphical displays. Often these receivers are built to only support the audio component of AirPlay, much like AirTunes.

Bluetooth devices (headsets, speakers) that support the A2DP profile also appear as AirPlay receivers when paired with an iOS device, although Bluetooth is a device-to-device protocol that does not rely on a wireless network access point.

During the January 2019 Consumer Electronics Show (CES) in Las Vegas, television makers Samsung, LG, Vizio, and Sony announced they would be producing sets with built-in AirPlay 2 receiving capability. LG announced that television models that are Airplay 2-enabled will include the 2019 OLED, NanoCell SM9X, UHD UM7X, and LG NanoCell SM8X models.

## Protocols

AirPlay and AirTunes work over Wi-Fi. Originally, devices had to be connected to the same Wi-Fi network, but since iOS 8 devices can use Wi-Fi Direct and thus does not require an existing Wi-Fi network.

The AirTunes part of the AirPlay protocol stack uses UDP for streaming audio and is based on the RTSP network control protocol. The streams are transcoded using the Apple Lossless codec with 44100 Hz and 2 channels symmetrically encrypted with AES, requiring the receiver to have access to the appropriate key to decrypt the streams. The stream is buffered for approximately 2 seconds before playback begins, resulting in a small delay before audio is output after starting an AirPlay stream.

The protocol supports metadata packets that determine the final output volume on the receiving end. This makes it possible to always send audio data unprocessed at its original full volume, preventing sound quality deterioration due to reduction in bit depth and thus sound quality which would otherwise occur if changes in volume were made to the source stream before transmitting. It also makes possible the streaming of one source to multiple targets each with its own volume control. The AirPort Express' streaming media capabilities use Apple's Remote Audio Output Protocol (RAOP), a proprietary variant of RTSP/RTP. Using WDS-bridging, the AirPort Express can allow AirPlay functionality (as well as Internet access, file and print sharing, etc.) across a larger distance in a mixed environment of wired and up to 10 wireless clients.

## AirPlay Mirroring

At WWDC 2011, Steve Jobs, then CEO of Apple Inc., announced AirPlay Mirroring as a feature in iOS 5 where the user can stream the screen from an iPad 2 to a HDTV wirelessly and securely without the need for cables.

AirPlay Mirroring is a slightly different technology that allows specific content to be broadcast from a variety of iOS devices and iTunes to an Apple TV (2nd Gen or later). The exact composition of the protocols that AirPlay Mirroring uses have not yet fully been discovered, or reverse-engineered. However, an unofficial AirPlay protocol specification is available. Supported hardware (when using OS X Mountain Lion or later) includes any 2011 or later iMac, Mac mini, MacBook Air, MacBook Pro, or the Mac Pro.

## Reverse Engineering

When the protocol was known as AirTunes, it was reverse-engineered by Jon Lech Johansen in 2008.

On April 8, 2011, James Laird reverse-engineered and released the private key used by the Apple AirPort Express to decrypt incoming audio streams. The release of this key means that third-party software and devices modified to use the key will be able to decrypt and play back or store AirPlay streams. Laird released ShairPort as an example of an audio-only software receiver implementation of AirPlay. Soon more followed and in 2012 the first AirPlay audio and video receiver for PC came with a product called AirServer.

# CHROMECAST

Chromecast is a line of digital media players developed by Google. The devices, designed as small dongles, enable users with a mobile device or personal computer to play Internet-streamed audio-visual content on a high-definition television or home audio system through mobile and web apps that support the Google Cast technology. Alternatively, content can be mirrored from the Google Chrome web browser running on a personal computer, as well as from the screen of some Android devices.

The first-generation Chromecast, a video streaming device, was announced on July 24, 2013, and made available for purchase on the same day in the United States for US$35. The second-generation Chromecast and an audio-only model called Chromecast Audio were released in September 2015. A model called Chromecast Ultra that supports 4K resolution and high dynamic range was released in November 2016. A third generation of the HD video Chromecast was released in October 2018.

Critics praised the Chromecast's simplicity and potential for future app support. The Google Cast SDK was released on February 3, 2014, allowing third parties to modify their software to work with Chromecast and other Cast receivers. According to Google, over 20,000 Google Cast–ready apps are available, as of May 2015. Over 30 million units have sold globally since launch, making the Chromecast the best-selling streaming device in the United States in 2014, according to NPD Group. From Chromecast's launch to May 2015, it handled more than 1.5 billion stream requests.

## Development

According to Google, the Chromecast was originally conceived by engineer Majd Bakar. His inspiration for the product came around 2008 after noticing the film-viewing tendencies of his wife Carla Hindie. Using her laptop, she would search for a film to watch on a streaming service and add it to her queue, before closing her laptop and using a gaming device to play the film on a television. She took these steps because she found television interfaces difficult to use to search for content. Bakar found the whole process inefficient and wanted to build a phone-based interface that would allow video to play on a large display through a small hardware device. After joining Google in 2011 to work on products that "would change how people used their TVs", Bakar pitched the idea for the Chromecast. Development on the product began in 2012; late that year, Bakar brought home a beta version of the product for Hindie to test. The device was launched in July 2013.

## Features and Operation

A first-generation Chromecast plugged into the HDMI port of a TV.

Chromecast offers two methods to stream content: the first employs mobile and web apps that support the Google Cast technology; the second allows mirroring of content from the web browser Google Chrome running on a personal computer, as well as content displayed on some Android devices. In both cases, playback is initiated through the "cast" button on the sender device.

When no content is streamed, video-capable Chromecasts display a user-personalizable content feed called "Backdrop" that can include featured and personal photos, artwork, weather, satellite images, weather forecasts, and news.

If a television's HDMI ports support the Consumer Electronics Control (CEC) feature, pressing the cast button will also result in the video-capable Chromecast automatically turning on the TV and switching the television's active audio/video input using the CEC command "One Touch Playback".

## Hardware and Design

Chromecast devices are dongles that are powered by connecting the device's micro-USB port to an external power supply or a USB port. Video-capable Chromecasts plug into the HDMI port of a high-definition television or monitor, while the audio-only model outputs sound through its integrated 3.5 millimeter audio jack/mini-TOSLINK socket. By default, Chromecasts connect to the Internet through a Wi-Fi connection to the user's local network; a standalone USB power supply with an Ethernet port, introduced in July 2015 for US$15, allows for a wired connection.

## First Generation

The first-generation video-capable Chromecast.

The original Chromecast measures 2.83 inches (72 mm) in length and has an HDMI plug built into the body. It contains the Marvell Armada 1500-mini 88DE3005 system on a chip running an ARM Cortex-A9 processor. The SoC includes codecs for hardware decoding of the VP8 and H.264 video compression formats. Radio communication is handled by AzureWave NH–387 Wi-Fi which implements 802.11 b/g/n (2.4 GHz). The device has 512 MB of Micron DDR3L RAM and 2 GB of flash storage.

## Second Generation

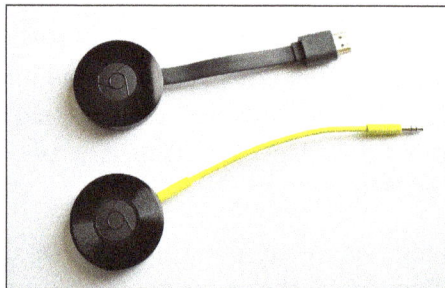

The second-generation video capable Chromecast and audio-only Chromecast Audio.

The second-generation Chromecast has a disc-shaped body with a short length of HDMI cable attached (as opposed to the HDMI plug built into the original model). The cable is flexible and can magnetically attach to the device body for more positioning options behind a television. The second-generation model uses a Marvell Armada 1500 Mini Plus 88DE3006 SoC, which has dual ARM Cortex-A7 processors running at 1.2 GHz. The unit contains an Avastar 88W8887, which has improved Wi-Fi performance and offers support for 802.11 ac and 5 GHz bands, while containing three adaptive antennas for better connections to home routers. The device contains 512 MB of Samsung DDR3L RAM and 256 MB of flash storage.

The model number NC2-6A5 may be a reference to the registry number "NCC-1701" of the fictional starship USS *Enterprise* from the *Star Trek* franchise, the "saucer section" of which the device resembles: $NC^2$ can be read as NCC, and 6A5 converted from hexadecimal is 1701.

## Third Generation

The third-generation Chromecast was launched in October 2018. It supports 1080p video at 60fps, but does not support 4K video.

## Chromecast Audio

A Chromecast Audio device connected to the auxiliary (AUX) port of a powered speaker.

Introduced in September 2015, Chromecast Audio is a variation of the second-generation Chromecast designed for use with audio streaming apps.

Chromecast Audio features a 3.5 millimeter audio jack/mini-TOSLINK socket, allowing the device to be attached to speakers and home audio systems. One side of the device is inscribed with circular grooves, resembling those of a vinyl record. A December 2015 update introduced support for high-resolution audio (24-bit/96 kHz) and multi-room playback; users can simultaneously play audio across multiple Chromecast Audio devices in different locations by grouping them together using the Google Home mobile app. The feature made Chromecast Audio a low-cost alternative to Sonos' multiple-room music systems.

With the advent of Google Home smart speakers, the device became tangential to Google's product strategy and was discontinued in January 2019. In addition, the third-generation Chromecast supports Chromecast Audio technology, allowing it to be paired with other devices for multi-room synchronized playback.

The model number RUX-J42 may have been a reference to the Jimi Hendrix albums *Are You Experienced* (stylized "R U eXperienced") and *Midnight Lightning*, which had the internal code J-42. Chromecast Audio was also developed with the internal code-name Hendrix.

## Chromecast Ultra

Chromecast Ultra is similar in design to the second-generation model, but features upgraded hardware that supports the streaming of 4K resolution content, as well as high-dynamic range through the HDR10 and Dolby Vision formats. Google stated that the Chromecast Ultra loads video 1.8 times faster than previous models. Unlike previous models that could be powered through a USB port, the Chromecast Ultra requires the use of the included power supply for connecting to a wall outlet. The power supply also offers an Ethernet port for a wired connection to accommodate the fast network speeds needed to stream 4K content.

## Software

### Google Cast SDK and Compatible Apps

Icon for the "cast button", which is used to connect, control and disconnect
from Google Cast receivers. The button can also represent compatible
non-Cast receivers, such as Bluetooth audio players.

At the time of Chromecast's launch, four compatible apps were available: YouTube and Netflix were supported as Android, iOS, and Chrome web apps; Google Play Music and Google Play Movies & TV were also supported, but originally only as Android apps. Additional Chromecast-enabled apps would require access to the Google Cast software development kit (SDK). The SDK was first released as a preview version on July 24, 2013. Google advised interested developers to use the SDK to create and test Chromecast-enabled apps, but not distribute them. While that admonition remained in force,

Chromecast-enabled applications for Hulu Plus and Pandora Radio were released in October 2013, and HBO Go in November. Google opened the SDK to all developers on February 3, 2014. In its introductory documentation and video presentation, Google said the SDK worked with both Chromecast devices and other unnamed "cast receiver devices". Chromecast product manager Rish Chandra said that Google used the intervening time to improve the SDK's reliability and accommodate those developers who sought a quick and easy way to cast a photo to a television without a lot of coding.

Over time, many more applications have been updated to support Chromecast. At Google I/O 2014, the company announced that 6,000 registered developers were working on 10,000 Google Cast–ready apps; by the following year's conference, the number of compatible apps had doubled. Google's official list of compatible apps and platforms is available on the Chromecast website. Google has published case studies documenting Chromecast integration by Comedy Central, Just Dance Now, Haystack TV and Fitnet.

In July 2019, the Amazon Prime apps for Android and iOS added Chromecast support, marking the first time Amazon's streaming service supported the device. The move followed a four-year dispute between Google and Amazon in which Amazon stopped selling Chromecast devices and Google pulled YouTube from Amazon Fire TV.

The development framework has two components: a sender app based on a vendor's existing Android or iOS mobile app, or desktop Web app, which provides users with content discovery and media controls; and a receiver app, executing in a Chrome browser-like environment resident on the cast receiver device. Both make use of APIs provided by the SDK.

## Device Discovery Protocols

Chromecast uses the mDNS (multicast Domain Name System) protocol to search for available devices on a Wi-Fi network. Chromecast previously used the DIAL (Discovery and Launch) protocol, co-developed by Netflix and YouTube.

## Operating System

At the introductory press conference, Mario Queiroz, Google's VP of Product Management, said that the first-generation Chromecast ran "a simplified version of Chrome OS." Subsequently, a team of hackers reported that the device is "more Android than ChromeOS" and appears to be adapted from software embedded in Google TV. As with Chrome OS devices, Chromecast operating system updates are downloaded automatically without notification.

## Mobile App

Chromecast is managed through the Google Home app, which enables users to set up new devices and configure existing ones (such as specifying which "Backdrop" images are shown when no other content is cast). Users can also search for streaming content that is available on installed Google Cast-enabled apps. The app manages other Google Cast-supported devices, including the Google Home smart speaker.

Originally called simply "Chromecast", the app was released concurrently with the original Chromecast video model and is available for both Android and iOS mobile devices. The app was released outside the US in October 2013.

In May 2016, the Chromecast app was renamed Google Cast due to the proliferation of non-Chromecast products that support casting. In October 2016, Google Cast was renamed Google Home, the name also given to the company's smart speaker—leaving "Google Cast" as the name of the technology.

## GOOGLE CAST

Google Cast, branded for consumer devices as Chromecast built-in, is a proprietary protocol developed by Google that enables mobile devices and personal computers to initiate and control playback of Internet-streamed audio/video content on a compatible device, such as a digital media player connected to a high-definition television or home audio system. The protocol was first launched on July 24, 2013, to support Google's first-generation Chromecast player. The Google Cast SDK was released on February 3, 2014, allowing third parties to modify their software to support the protocol. According to Google, over 20,000 Google Cast-ready apps were available as of May 2015. Google Cast would later be built into the Nexus Player and other Android TV devices (such as televisions), as well as soundbars, speakers, and subsequent Chromecast players.

### Operation

Google Cast receivers can stream content via two methods: the first employs mobile and web apps that support the Google Cast technology; the second allows mirroring of content from the web browser Google Chrome running on a personal computer, as well as content displayed on some Android devices. In both cases, playback is initiated through the "cast" button on the sender device.

- The primary method of playing media on the device is through Google Cast–enabled mobile and web apps, which control program selection, playback, and volume. Google Cast receiver devices stream the media from the web within a local version of the Chrome browser, thus freeing the sender device up for other tasks, such as answering a call or using another application, without disrupting playback. Mobile apps enabled for Google Cast are available for both Android 4.1+ and iOS 7.0+; web apps enabled for Google Cast are available on computers running Google Chrome (on Microsoft Windows 7+, macOS 10.7+, and Chrome

OS for Chromebooks running Chrome 28+) through the installation of the "Cast extension" in the browser. Streamed content can be Internet-based, as provided by specific apps, or reside on the sender device's local storage. Apps that provide access to the latter include AllCast, Avia, Plex, and Google Photos.

- Content can also be mirrored from a tab of the Chrome browser (with the Cast extension) on a personal computer or from the screen of some Android 4.4+ devices. In the case of "tab casting", the quality of the image depends on the processing power of the device, and minimum system requirements apply to video streaming. Content that uses plug-ins, such as Silverlight and QuickTime, does not fully work, as the stream may lack sound or image. Similarly, screen images mirrored from Android devices are typically degraded, reflecting the fact that video displayed on the smaller screens of tablets and smartphones is usually downscaled.

Sender devices previously needed to be connected to the same Wi-Fi network as a Google Cast receiver device to cast content, until the addition of a "guest mode" feature on December 10, 2014; When enabled, the feature allows sender devices to discover a nearby player by detecting ultrasonic audio emitted by the television or speaker system to which the player is connected; alternatively, the sender device can be paired with the receiver device using a four-digit PIN code. Guest mode is only available for Chromecasts; the Nexus Player and Android TV devices do not support the feature.

## Framework

The development framework has two components: a sender app and a receiver app, both of which make use of APIs provided by the SDK.

- The sender app is based on a vendor's existing Android or iOS mobile app, or desktop web app, and provides users with content discovery and media controls, including the ability to select to which device content is streamed. Under the hood, sender apps can detect receiver devices on the same local network, establish a secure channel, and exchange messages.

- The receiver app is a web app executing in a Chrome browser-like environment resident on the cast receiver device. Receiver apps of varying complexities can be developed depending on the variety of content formats the app can play. For example, a simple receiver app might just play HTML5 content, whereas custom receiver apps, which require more programming effort, can take a variety of streaming protocols, including MPEG-DASH, HTTP Live Streaming (HLS), and the Microsoft Smooth Streaming Protocol.

## Supported Media

Chromecast supports the image formats BMP, GIF, JPEG, PNG, and WEBP, with a display size limitation of 720p (1280 × 720 pixels). Supported audio codecs are

HE-AAC, LC-AAC, MP3, Vorbis, WAV (LPCM), FLAC (up to 96kHz/24-bit) and Opus; AC-3 (Dolby Digital) and E-AC-3 (EC-3, Dolby Digital Plus) are available for audio pass-through. Supported video codecs for the first and second generation Chromecast are H.264 High Profile Level 4.1 (decoding up to 720p/60 frames per second (fps) or 1080p/30fps) and VP8. Supported video codecs for the third generation Chromecast are H.264 High Profile Level 4.2 (decoding up to 720p/60 (fps) or 1080p/60fps) and VP8. The supported video codecs for the Chromecast Ultra are HEVC / H.265 Main and Main 10 Profiles up to level 5.1 (2160p/60fps) and VP9 Profile 0 and Profile 2 up to level 5.1 (2160p/60fps).

## Additional Functionality and APIs

At International CES 2015, Google announced an expansion to Google Cast called "Google Cast for audio", which allows apps that support the Google Cast SDK to play audio through compatible Wi-Fi–connected speakers, soundbars, and receivers. Manufacturers supporting Google Cast as a built-in function in their speakers include LG and Sony.

In May 2015, Google introduced new sets of APIs to Google Cast. The Cast Remote Display APIs allow developers to create second-screen experiences for apps such as games without needing to mirror displays. The Game Manager APIs offer developers more options for creating multiplayer games. Lastly, additional APIs were provided to control autoplaying and queuing of content.

In September 2015, Google announced "Fast Play" and accompanying developer tools, which are aimed at reducing the delays between loading content. In a typical scenario, if a user viewed the first three episodes of a television series, the fourth episode might load in the background. The feature's release has since been delayed.

## EZCAST

EZCast is a line of digital media players, built by Actions Microelectronics, that allows users to mirror media content from smart devices, including mobile devices, personal computers, and project to high-definition televisions.

The first generation of EZCast was developed in 2013, shipped 1 million units within a year, and accumulated more than 2 million EZCast app users worldwide. The latest device in the family, called EZCast 4K, was launched in November 2016 which supports 4K HEVC video streaming.

EZCast technology is built into a dongle that interacts with EZCast app to stream content from smart devices, and it works across Android, Chrome OS, iOS, macOS, Windows and Windows Phone.

EZCast SDK has been released to enable third party development on Android and iOS. In 2018 became possible to voice control EZCast 2 and EZCast 4K devices using Google Assistant.

# MIRACAST

Miracast is a standard for wireless connections from devices (such as laptops, tablets, or smartphones) to displays (such as TVs, monitors, or projectors), introduced in 2012 by the Wi-Fi Alliance. It can roughly be described as "HDMI over Wi-Fi", replacing the cable from the device to the display.

The Wi-Fi Alliance launched the Miracast certification program at the end of 2012. Devices that are Miracast-certified can communicate with each other, regardless of manufacturer. Adapters became available that may be plugged either into HDMI or USB ports, allowing devices without built-in Miracast support to connect via Miracast.In 2013, Nvidia announced support for Miracast.

Miracast employs the peer-to-peer Wi-Fi Direct standard. It allows sending up to 1080p HD video (H.264 codec) and 5.1 surround sound (AAC and AC3 are optional codecs, mandated codec is linear pulse-code modulation – 16 bits 48 kHz 2 channels). The connection is created via WPS and therefore is secured with WPA2. IPv4 is used on the Internet layer. On the transport layer, TCP or UDP are used. On the application layer, the stream is initiated and controlled via RTSP, RTP for the data transfer.

## Devices

The Wi-Fi Alliance maintains a list of certified devices, which numbered over 6,700 as of 9 March 2017.

Nvidia announced support in 2012 for their Tegra 3 platform, and Freescale Semiconductor, Texas Instruments, Qualcomm, Marvell Technology Group and other chip vendors have also announced their plans to support it. Actiontec Electronics also supports Miracast with its line of ScreenBeam products.

Both devices (the sender and the receiver) need to be Miracast certified for the technology to work. However, to stream music and movies to a non-certified device, Miracast adapters are available that plug into HDMI or USB ports.

On 29 October 2012, Google announced that Android version 4.2+ (from updated version of Jelly Bean) are supporting the Miracast wireless display standard, and by default have it integrated. With Android 6.0 Marshmallow, released in 2015, Miracast support was dropped.

As of 8 January 2013, the LG Nexus 4 and Sony's Xperia Z, ZL, T and V officially supported the function, as did HTC One, Motorola in their Droid Maxx & Ultra flagships, and Samsung in its Galaxy S III and Galaxy Note II under the moniker *AllShare Cast*. The Galaxy S4 uses Samsung Link for its implementation.

In October 2013, BlackBerry released its 10.2.1 update to most of the existing BlackBerry 10 devices available at that time. As of March 2015, the BlackBerry Q10, Q5, Z30, and later models support Miracast streaming; the BlackBerry Z10 does not support Miracast, due to hardware limitations.

In April 2013, Rockchip unveiled a Miracast adapter powered by the RK2928.

Microsoft also added support for Miracast in Windows 8.1 (announced in June 2013) and Windows 10. This functionality first became available in the Windows 8.1 Preview, and is available on hardware with supported Miracast drivers from hardware (GPU) manufacturers such as those listed above.

The WDTV Live Streaming Media Player added Miracast support with firmware version 2.02.32

The Amazon Fire TV Stick, which started shipping on 19 November 2014, also supports Miracast.

The Roku streaming stick and Roku TV started providing support for Miracast starting October 2014.

On 28 July 2013, Google announced the availability of the Chromecast powered by a Marvell DE3005-A1, but despite the similarity in name and Google's early support of Miracast in Android, the Chromecast does not support Miracast.

As of late April 2016, the Ubuntu Touch-powered Meizu Pro 5 supported Miracast in OTA-11.

## Functionality

The technology was promoted to work across devices, regardless of brand. Miracast devices negotiate settings for each connection, which simplifies the process for the users. In particular, it obviates having to worry about format or codec details. Miracast is "effectively a wireless HDMI cable, copying everything from one screen to another using the H.264 codec and its own digital rights management (DRM) layer emulating the HDMI system". The Wi-Fi Alliance suggested that Miracast could also be used by a set-top box wanting to stream content to a TV or tablet.

## Types of Media Streamed

Miracast can stream videos that are in 1080p, media with DRM such as DVDs, as well as protected premium content streaming, enabling devices to stream feature films and other copy-protected materials. This is accomplished by using a Wi-Fi version of the same trusted content mechanisms used on cable-based HDMI and DisplayPort connections.

## Display Resolution

- 27 Consumer Electronics Association (CEA) formats, from 640 x 480 up to 4096 x 2160 pixels, and from 24 to 60 frames per second (fps).

- 34 Video Electronics Standards Association (VESA) formats, from 800 x 600 up to 2560 x 1600 pixels, and from 30 to 60 fps.

- 12 handheld formats, from 640 x 360 up to 960 x 540 pixels, and from 30 to 60 fps.

- Mandatory: 1280 x 720p30 (HD).

- Optional: 3840 x 2160p60 (4K Ultra HD).

## Video

Mandatory: ITU-T H.264 (Advanced Video Coding [AVC]) for HD and Ultra HD video; supports several profiles in transcoding and non-transcoding modes, including Constrained Baseline Profile (CBP), at levels ranging from 3.1 to 5.2.

Optional: ITU-T H.265 (High Efficiency Video Coding [HEVC]) for HD and Ultra HD video; supports several profiles in transcoding and non-transcoding modes, including Main Profile, Main 444, SCC-8 bit 444, Main 444 10, at levels ranging from 3.1 to 5.1.

## Audio

Mandated audio codec: Linear Pulse-Code Modulation (LPCM) 16 bits, 48 kHz sampling, 2 channels.

Optional audio codecs, including:

- LPCM mode 16 bits, 44.1 kHz sampling, 2 channels,

- Advanced Audio Coding (AAC) modes,

- Dolby Advanced Codec 3 (AC3) modes,

- E-AC-3,

- Dolby TrueHD, Dolby MAT modes,

- DTS-HD mode,

- MPEG-4 AAC and MPEG-H 3D Audio modes,

- AAC-ELDv2.

## OPENFLINT

OpenFlint is an open technology used for displaying ("casting") content from one computerized device on the display of another. Usually this would be from a smaller personal device (like a smartphone) to a device with a larger screen suitable for viewing by multiple spectators (like a TV).

Development of OpenFlint was initiated in 2014 by the Matchstick project, which is a crowd-funded effort to create a miniature piece of hardware suitable for running an OpenFlint server casting to a screen through an HDMI connection. This is similar in concept to Google's Chromecast device that uses Google Cast.

The Matchstick TV devices are powered by Firefox OS, but as an open technology OpenFlint itself is not tied to any specific operating system or hardware.

As of July 2015, no consumer-grade OpenFlint-enabled products have shipped, but Matchstick developer devices have been shipping since late 2014, and the first round of devices for backers of the Matchstick Kickstarter project were expected to ship in February 2015, but were delayed until August 2015.

A demonstration OpenFlint server can be set up on an ordinary laptop or desktop computer running Linux by following instructions.

The Matchstick TV dongle project was canceled due to issues implementing DRM into Firefox OS.

## WIRELESS HDMI TRANSMISSION PROTOCOLS

### WHDI

As the first practical and successful wireless HD transmission protocol, WHDI can deliver wireless interactive HD video from any device to any displayer.

WHDI has changed our previous impression of wireless video: lose of quality, long latency, distortion of image. It provides us uncompressed HD signal up to 1080p resolution in the 5Ghz unlicensed band, with support for 3D TV and 5.1 surrounding sound as well.

What WHDI gives customers is the convenience and the flexibility. 5GHz spectrum regulations are naturally superior for wireless transmission. The range under this WHDI wireless protocol is beyond 100 feet. Signal can easily penetrate brick walls. You can easily setup a whole wireless entertainment system without any worrying about the location of devices, whether it is for the source or the playback .

Any Wireless HDMI device with this protocol comes with near zero and neglectable latency, namely less than 1ms. WHDI also features feedback controls channel, so you can control the source device through its own remote or gaming paddle. These make it a perfect partner for a gaming console and any circumstance requiring harmony of separate screens.

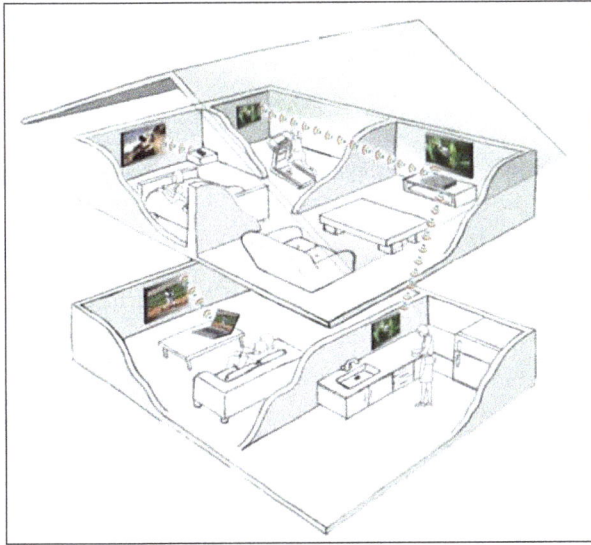

You can easily connect any video sources through a WHDI Wireless HDMI transmitting kit with HD displayers wirelessly. These devices include: PC, Notebooks, Smart Phones, Satellite boxes, Xbox and Blu-ray players.

In short terms, WHDI equals uncompressed 1080p+ low latency+ multi-room availability + low power consumption. It surpasses competitive technologies by providing outstanding and consistent picture quality equivalent to wired HDMI cable.

## WHDI Wireless HDMI Devices

As the most successful wireless HD protocol. Many companies choose WHDI for their products.

## Nyrius ARIES Pro Digital Wireless HDMI Transmitter

The Nyrius ARIES Pro digital wireless HDMI transmitter is designed to meet the increasing demanding of high quality video transmission on the market. This Nyrius wireless HDMI transmitter and receiver system integrates WHDI technology, given by an Israeli Hi-tech company AMIMON, for truly HD wireless video broadcast.

## IOGEAR GW3DHDKIT Wireless 3D Digital Kit

This IOGEAR GW3DHDKIT Wireless 3D Digital Kit is one of the best wireless HDMI transmitter&receiver kits. It will give you the uncompressed wireless full-rate HD audio and video up to 100 feet between the transmitter and the receiver, which can be mounted on or behind your wall mounted HDTV.

## Atlona LinkCast Wireless HD Audio/Video Station

Atlona LinkCast Wireless HD system Atlona LinkCast Wireless HDMI Audio/Video System is a highly rewarded wireless transmission solution, which had won the Electronic House 2012 "Product of the Year" award and 2012 CES Innovations Award.

## WirelessHD: Powerful and Promising WirelessHD Protocol

The WirelessHD specification is based on a 7 GHz channel in the 60 GHz Extremely High Frequency radio band. Unlike other wireless HD signal broadcast protocol WirelessHD choose a high frequency radio realm 60 GHZ to fulfill its mission.

All characteristics of this specific wireless communication consortium are determined by this unusual decision. This fundamental difference of 60 GHz systems has some technical advantages compared to other network protocols like Wi-Fi, but also has some inherent limitations.

### Pros of WirelessHD

WirelessHD protocol utilizes this very high frequency to increase the amount of network bandwidth and effective data rates they can support. It can provide uncompressed digital transmission of HD video and audio and data signals, essentially making it equivalent of a wireless HDMI. The 1.0 version of this standard supported data rates of 4 Gbps, while new version 1.1 increases the maximum data rate up to 28 Gbit/s. This is the highest theoretical data rate of all wireless HD transmission nowadays. So WirelessHD supports 1080p, 3D formats and 4K resolution without any problem.

Another superior point of WirelessHD is the compatibility of high rates wireless audio formats The DVDO Air Based WirelessHD protocol is the only available wireless HDMI product can transmit 1080p video and 7.1-channel audio simultaneously, While WHDI product like Nyrius Pro are now capable of 5.1 channel, without using an HDMI cable.

Like WHDI it also features remote controlling which enables users to control the WirelessHD devices and choose between sources for the display. The WirelessHD specification supports DTCP(Digital Transmission Content Protection) for encrypted content with copyright.

## Cons of WirelessHD

High frequencies equal short waves. These waves are might likely to reflect on any smooth surfaces. Signals under this protocol are not able to pass through most physical obstructions. Although WirelessHD try to enlarge the beaming range by increasing the signal's effective radiated power and utilizing wall reflections. This typical 60 Gbps HDMI wireless protocol vivid connection is still within the distances of 30 feet.

## DVDO WirelessHD System

This wireless DVDO AIR system works to transfer full, uncompressed HD signal to any HD display device, so you can watch 1080p movies and enjoy full 7.1 channel surround sound wirelessly. Unlike other wireless HDMI streaming devices DVDO Air wireless HD connection system works under The Wireless HD (also called WiHD) standard, which runs on 60 GHZ realm instead of the normally crowded 5 GHZ.

WiGig is the trademark for Wireless Gigabit Alliance, an alliance formed to develop and promote the wireless Wi-Fi transmission protocol 802.11ad. Like WirelessHD, this protocol is operating over the unlicensed 60 GHZ frequency band. The WiGig specification allows devices to communicate without wires at multi-gigabit speeds. Theoretically a typical WiGig device can deliver data transfer rates up to 7 Gbit/s with whole 7 GHZ bandwidth.It is approximately as fast as an 8 antenna 802.11ac transmissions. If divided to four 2.16-GHz wide channels, one single-channel can archive a speed up to 4.6 Gbit/s.

WiGig Alliance has merged with the Wi-Fi Alliance and its wireless HDMI standard is defined in the IEEE 802.11ad. This means WiGig get a full compatibility with existing Wi-Fi IEEE protocols. Most consumers are well aware of IEEE protocols because they have experience with all the earlier ones, namely IEEE 802.11a, b, g, and n. The 11b/g/n standards operate in the 2.4-GHz band. The 11a/ac/n standards operate in the 5-GHz band, and the 802.11ad standard targets the 60-GHz band. The WiGig tri-band Wi-Fi chips can switch seamlessly from one standard to another in one of the three Wi-Fi bands: 2.4, 5, or 60 GHz.

Maybe WirelessHD has a step in advance for giving of tangible products. But WiGig will not lose this battle lightly. The 60 GHz signal cannot typically penetrate walls connections through walls. WiGig high rates wireless HDMI connections are also restricted to a single room. Nonetheless, unlike WirelessHD when roaming away from the main room WiGig can switch to make use of the other lower bands,2.4 or 5 GHZ, at a much lower rate, but which propagate through walls.

WiGig and WirelessHD are widely perceived as competing technologies. Some believe WiGig may even replace Wi-Fi technology someday, although this would require solving its range limitation issues. There are no practical WiGig product except Dell has made some trying.

## Other Potential Wireless HDMI Transmission Methods

Miracast is a Wi-Fi direct peer-to-peer connection without interaction of wireless router. In this respect, it functions like a Bluetooth connection. Miracast supports 1080p HD video and 5.1 surround sound as well. It allows users to mirror the display from a phone or tablet onto a TV, share a laptop screen with the conference room projector simultaneously.The most common used product are Google Chromecast.

Intel has developed its own wireless transmission protocol: Wireless Display(WiDi). Like all the standards referred to above, WiDi also supports HD 1080p resolution and 5.1 channeled audio. Low latency and interacting app are featured with this wireless transmission technology. It is more for source devices like tablets and smart phones. Laptops/ Ultrabooks with WiDi 3.5 and higher are Miracast-compatible. There are plenty of Wireless Display Dongles and Laptops supporting WiDi transmission.

AirPlay is the most famous licensed proprietary protocol by Apple. Originally it is just for Apple's own software and devices to stream audio, video, and photos. In order to be competitive with others, Apple has licensed AirPlay to third-party company. Many products are now compatible for AirPlay,but the most successful AirPlay device is no doubt the Apple TV by Apple inc. itself.

# Orthogonal Frequency Division Multiplexing

Orthogonal frequency division multiplexing is used for wideband digital communication through encoding digital data on multiple carrier frequencies. It has varied applications in digital television and audio broadcasting, wireless networks, power line networks, and 4G mobile communications. This chapter discusses orthogonal frequency division multiplexing in detail.

The OFDM system is a twin of the Frequency Division multiplexing method. OFDM practices the similar rule as of FDM wherever many messages unit to be sent through one radio channel in associate unionized manner. FDM has various frequencies for each modulation station. Each and every frequency signal has ample frequency difference between them that they don't overlap to every alternative within the frequency spectrum. Every frequency signal is on an individual basis sift through a band pass filter to require the opposite complete signal excluding the required signal that is base station considering for at the receiver. The acknowledged signal is converse to get back the first signal.

By means that of OFDM technology transmission of information follows through an enormous range of information measure carriers. All the carriers unit gapped oftentimes by frequency that generates a block of spectrum. All the carriers unit orthogonally formed by the frequency gap in addition to time synchronization. These carriers are generated as they are doing not conclude up to interference within the frequency spectrum.

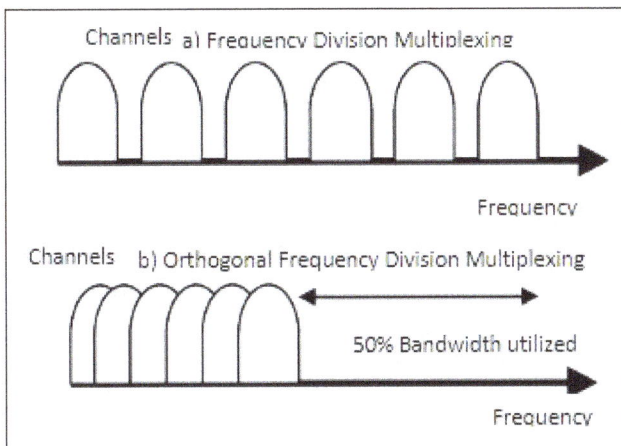

OFDM Signal.

Figure displays the difference amongst the standard non-overlapping multicarrier technique and also the overlapping multi-carrier modulation technique. It displays the five hundred information measure improved by mistreatment multicarrier modulation technique.

OFDM, furthermore known as multicarrier modulation, practices a couple of carrier signals alike frequency, sending little number of bits on the each channel. This is comparable to FDM. Nevertheless, in the case of OFDM, all the sub-channels are devoted to single records source.

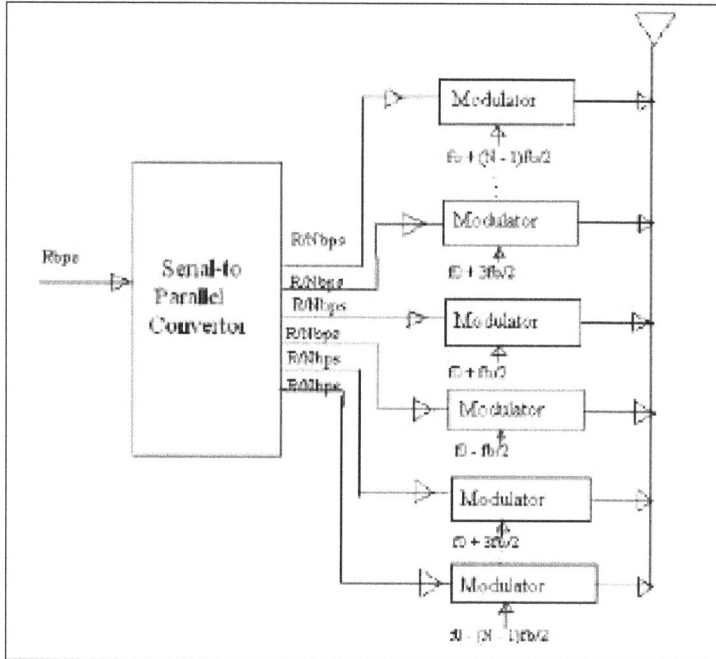

Orthogonal Frequency Division Multiplexing.

Figure illustrates OFDM. Assume there is an information circulate working at R bps and to have a bandwidth of $Nf_b$, targeted at $f_o$. The complete bandwidth is probably utilized to ship the records stream, so in that condition each bit duration would be 1/R. the synthetic is to break up the statistics circulation into N substreams, through a serial to parallel converter. Every sub carrier has a statistics data rate of R/N bps and is transmitted through complete subcarrier, among nearby sub-carriers of frequency. Now the bit length is N/R. To advantage a clearer know-how of OFDM, let us don't forget the scheme in phrases of their base frequency, $f_b$. This is the lowest frequency subcarrier. All the remaining subcarriers are integer multiples of the lowest frequency, particularly $2f_b$, $3f_b$ and so on, as proven in figure. OFDM device practices superior virtual signal processing techniques to allocate the facts over multiple providers at precise frequencies. The exact dating amid the subcarriers is devoted to as orthogonality. The final results, shown in determine three, is that the peaks of the strength spectral density of every sub-carrier appear at a factor at which the power of different sub-carriers is 0. With OFDM, the subcarriers are stuffed firmly collectively due to the fact there is negligible interference between adjoining subcarriers.

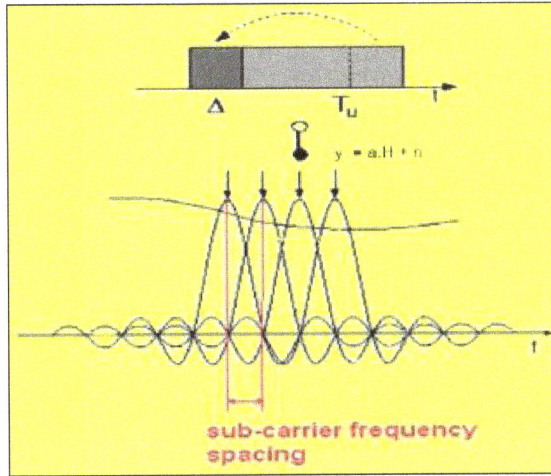

OFDM and the Orthogonality principle.

Figure depicts the set of OFDM subcarriers during a waveband starting with the bottom frequency. For transmission, the set of OFDM subcarriers is additional modulated to the next waveband. Additional vital, OFDM overcome international intelligence agency during a multipath atmosphere. International intelligence agency takes a bigger influence at larger bit rates, since the gap between bits, is reduced. With OFDM, the info or data rate is condensed by an element of N that will increase the symbol time by an element of N. Therefore, if the symbol amount is T; for the supply surge, the amount for the OFDM is National Trust. This intensely decreases the result of international intelligence agency. As a style concept, N is chosen in order that NTs is significantly bigger than the rootmean-square delay unfold of the complete channel.

## OFDM Transceiver

At the spell while a source transmits the facts it's far within the form of serial statistics circulate. This serial information movement takes to transfer the data into parallel form. As in OFDM method, each symbol typically communicates 40-4000 bits. Facts assigned to each symbol accounts on the modulation method used and the dimensions of subcarriers. As an instance, each subcarrier includes four bits of data for a subcarrier modulation system of 16-QAM and for the concern of transmission each symbol could be four hundred bits by one hundred subcarriers.

Figure displays the diagram of the OFDM transceiver. The subcarriers that don't seem to be transmission any knowledge take the worth of zero. All of the signals are there within the frequency domain. All such signals got to be wrapped into the time beforehand it will transmit. IFFT technique is used to change these signals into the time domain. All dissimilar samples of the IFFT states to one sub-carrier, at one time dynamical them into time domain. Nearly all of those subcarriers changed with knowledge.

The outer subcarriers don't seem to be altered with knowledge and have the amplitude rate of 0. These 0 amplitude subcarriers work as a guard band near the Nyquist frequency and competently work as an interpolation of signal and licenses for a true recite within the analogue anti-aliasing reconstruction filters. Because it is believed before the foremost vital deserves is victimization OFDM is lustiness of the multi-path delay unfold and is achieved by separating the input stream into N subcarriers. Currently the symbol length is explicit by N times smaller than reduce relative multipath delay unfold. Guard time is recognized to each symbol to rip off inter-symbol interference.

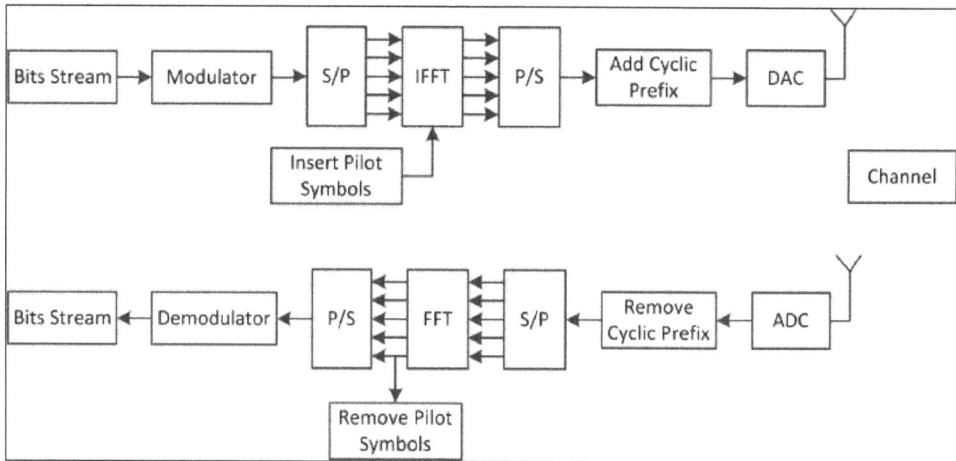

Block Diagram of OFDM Transceiver.

By investigation an additional guard amount at the commencement of each symbol will reduce the consequence of ISI within the OFDM. It's a revenant copy that rises the dimensions of the symbol wave shape. Every subcarriers of the information symbol unit has the definite size of cycles. Thanks to the aim of symbol end to end inserting copies, so the signal may be a non-stop signal while not breaks within the joins. OFDM method makes the symbol time better by copying to the end of a symbol and enhances them to the start of the next symbol. To receive the anticipated guard time is embarked on and FFT technique is applied at the receiver finish.

## OFDM Generation

It is significant to save the orthogonality of the carriers through careful control of the signal in order to generate the OFDM. To make this occur firstly OFDM requires the spectrum conferring to the input and modulation technique in scheme that is to be used. Each carrier is stated by the transmission of the data that is to be made. The essential carrier amplitude and phase of the carrier are then calculated. Conferring to the modulation system, required carrier amplitude and phase of the carrier is calculated generally using different techniques like BPSK, QPSK, or QAM. By using Inverse Fast Fourier Transform, required spectrum is now converted again to its time spectrum.

FFT coverts a cyclic time spectrum into frequency spectrum. This is completed by it. It is essential to discover the equal waveform as frequency signal shaped by a sum of the orthogonal sub parts. The amplitude of signal and phase of the sinusoidal parts displays the frequency of the time domain signal. The IFFT ensures that the opposite process, changes a frequency signal into time domain signal. An IFFT alters the length of complex data points as the size that is a power of 2, into the time domain with the distinguishable length of points.

Essential quantities of the OFDM transmitter and receiver are displayed within the figure. On this technique the signal, that is created first is a baseband signal and then the filtration of the signal is accomplished, after which paced up in frequency earlier than transmitting the signal.

## Addition of a Guard Period to OFDM

The power of OFDM is increased by the addition of a guard amount amongst the transmitted signals. Guard amount permits time for multipath signals from the previous symbol to bit by bit fade before the data from the present symbol is get along. Cyclic addition is that the best important guard amount to figure. Cyclic extension is important to rewrite the symbol by suggests that of the FFT. This contents the multipath resistance and also the time synchronization of the symbol.

Receiver accepts the signal from all the parts of the input, which since FFT in energy cares about general obtainable power is delivered to the decoder. Once the delay unfold is superior to the guard period, they begin to supply put down symbol interference. What is more, equipped echoes are sufficiently slight they are doing not turn out additional issues. It's same that utmost of the time multipath echoes are late to unfold superior than the guard amount is for the explanation that of reflection by various distant matters.

## Interference

Several components of the transmitted signal is set up with the aid of the receiver at sure instances inside the multipath ecosystem. This is the purpose that several multiple paths exist amongst the transmitter and receiver. This finish into the time spreading stretched, a specific received symbol into one penetrating into it. ISI is the period named for the overlapping of the signal. It in addition is a major intention in timing stability. Inter Carrier Interference (ICI) is the opposite gadget of interference.

## Cyclic Extension of OFDM Symbol

Time domain OFDM is often prolonged with the intention to keep away from the result of time spreading. With a purpose to elude ISI the size of cyclic prefix has to head overhead the most excess delay of the channel. Figure shows this perception. It also presents how the cyclic prefix evades the ISI.

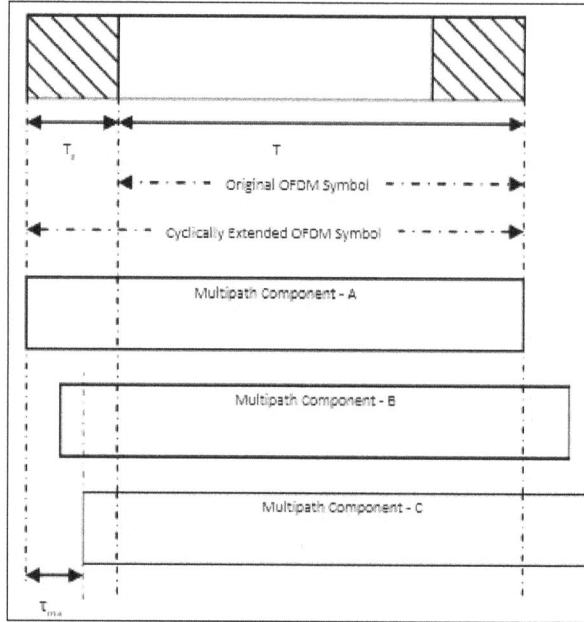

Cyclic Prefix Extension.

There are some important reasons to apply cyclic prefix:

- In an effort to hold the receiver time domain synchronization for the reason that an extended duration of silence can produce synchronization to be vanished.

- So as to modify the linear convolution of the signals.

- as well as channel to a round channel.

- It's far a ways comfy to generate in FPGAs.

## Characteristics and Principles of Operation

## Orthogonality

Conceptually, OFDM is a specialized frequency-division multiplexing (FDM) method, with the additional constraint that all subcarrier signals within a communication channel are orthogonal to one another.

In OFDM, the subcarrier frequencies are chosen so that the subcarriers are orthogonal to each other, meaning that cross-talk between the sub-channels is eliminated and inter-carrier guard bands are not required. This greatly simplifies the design of both the transmitter and the receiver; unlike conventional FDM, a separate filter for each sub-channel is not required.

The orthogonality requires that the subcarrier spacing is $Äf = \dfrac{k}{T_U}$ Hertz, where $T_U$ seconds is the useful symbol duration (the receiver-side window size), and $k$ is a positive integer, typically equal to 1. This stipulates that each carrier frequency undergoes $k$ more complete cycles per symbol period than the previous carrier. Therefore, with $N$ subcarriers, the total passband bandwidth will be $B \approx N \cdot \Delta f$ (Hz).

The orthogonality also allows high spectral efficiency, with a total symbol rate near the Nyquist rate for the equivalent baseband signal (i.e. near half the Nyquist rate for the double-side band physical passband signal). Almost the whole available frequency band can be used. OFDM generally has a nearly 'white' spectrum, giving it benign electromagnetic interference properties with respect to other co-channel users.

A simple example: A useful symbol duration $T_U$ = 1 ms would require a subcarrier spacing of $\ddot{A}f = \dfrac{1}{1ms} = 1\,kHz$ (or an integer multiple of that) for orthogonality. $N$ = 1,000 subcarriers would result in a total passband bandwidth of $N\Delta f$ = 1 MHz. For this symbol time, the required bandwidth in theory according to Nyquist is $BW = R\,/\,2 = (N\,/\,T_U)\,/\,2 = 0.5\,\text{MHz}$ (half of the achieved bandwidth required by our scheme), where $R$ is the bit rate and where $N$ = 1,000 samples per symbol by FFT. If a guard interval is applied, Nyquist bandwidth requirement would be even lower. The FFT would result in $N$ = 1,000 samples per symbol. If no guard interval was applied, this would result in a base band complex valued signal with a sample rate of 1 MHz, which would require a baseband bandwidth of 0.5 MHz according to Nyquist. However, the passband RF signal is produced by multiplying the baseband signal with a carrier waveform (i.e., double-sideband quadrature amplitude-modulation) resulting in a passband bandwidth of 1 MHz. A single-side band (SSB) or vestigial sideband (VSB) modulation scheme would achieve almost half that bandwidth for the same symbol rate (i.e., twice as high spectral efficiency for the same symbol alphabet length). It is however more sensitive to multipath interference.

OFDM requires very accurate frequency synchronization between the receiver and the transmitter; with frequency deviation the subcarriers will no longer be orthogonal, causing inter-carrier interference (ICI) (i.e., cross-talk between the subcarriers). Frequency offsets are typically caused by mismatched transmitter and receiver oscillators, or by Doppler shift due to movement. While Doppler shift alone may be compensated for by the receiver, the situation is worsened when combined with multipath, as reflections will appear at various frequency offsets, which is much harder to correct. This effect typically worsens as speed increases, and is an important factor limiting the use of OFDM in high-speed vehicles. In order to mitigate ICI in such scenarios, one can shape each subcarrier in order to minimize the interference resulting in a non-orthogonal subcarriers overlapping. For example, a low-complexity scheme referred to as WCP-OFDM (Weighted Cyclic Prefix Orthogonal Frequency-Division Multiplexing) consists of using short filters at the transmitter output in order to perform a potentially non-rectangular pulse shaping and a near perfect reconstruction using a single-tap per subcarrier equalization. Other ICI suppression techniques usually increase drastically the receiver complexity.

## Implementation using the FFT Algorithm

The orthogonality allows for efficient modulator and demodulator implementation using the FFT algorithm on the receiver side, and inverse FFT on the sender side. Although the principles and some of the benefits have been known since the 1960s, OFDM is popular for wideband communications today by way of low-cost digital signal processing components that can efficiently calculate the FFT.

The time to compute the inverse-FFT or FFT transform has to take less than the time for each symbol, which for example for DVB-T (FFT 8k) means the computation has to be done in 896 μs or less.

For an 8192-point FFT this may be approximated to:

$$\text{MIPS} = \frac{\text{computational complexity}}{T_{\text{symbol}}} \times 1.3 \times 10^{-6}$$

$$= \frac{147\ 456 \times 2}{896 \times 10^{-6}} \times 1.3 \times 10^{-6}$$

$$= 428$$

MIPS = Million instructions per second

The computational demand approximately scales linearly with FFT size so a double size FFT needs double the amount of time and vice versa. As a comparison an Intel Pentium III CPU at 1.266 GHz is able to calculate a 8192 point FFT in 576 μs using FFTW. Intel Pentium M at 1.6 GHz does it in 387 μs. Intel Core Duo at 3.0 GHz does it in 96.8 μs.

## Guard Interval for Elimination of Intersymbol Interference

One key principle of OFDM is that since low symbol rate modulation schemes (i.e., where the symbols are relatively long compared to the channel time characteristics) suffer less from intersymbol interference caused by multipath propagation, it is advantageous to transmit a number of low-rate streams in parallel instead of a single high-rate stream. Since the duration of each symbol is long, it is feasible to insert a guard interval between the OFDM symbols, thus eliminating the intersymbol interference.

The guard interval also eliminates the need for a pulse-shaping filter, and it reduces the sensitivity to time synchronization problems.

A simple example: If one sends a million symbols per second using conventional single-carrier modulation over a wireless channel, then the duration of each symbol would be one microsecond or less. This imposes severe constraints on synchronization and necessitates the removal of multipath interference. If the same million symbols per second are spread among one thousand sub-channels, the duration of each symbol can be longer by a factor of a thousand (i.e., one millisecond) for orthogonality with approximately the same bandwidth. Assume that a guard interval of 1/8 of the symbol length is inserted between each symbol. Intersymbol interference can be avoided if the multipath time-spreading (the time between the reception of the first and the last echo) is shorter than the guard interval (i.e., 125 microseconds). This corresponds to a maximum difference of 37.5 kilometers between the lengths of the paths.

The cyclic prefix, which is transmitted during the guard interval, consists of the end of the OFDM symbol copied into the guard interval, and the guard interval is transmitted followed by the OFDM symbol. The reason that the guard interval consists of a copy of the end of the OFDM symbol is so that the receiver will integrate over an integer number of sinusoid cycles for each of the multipaths when it performs OFDM demodulation with the FFT.

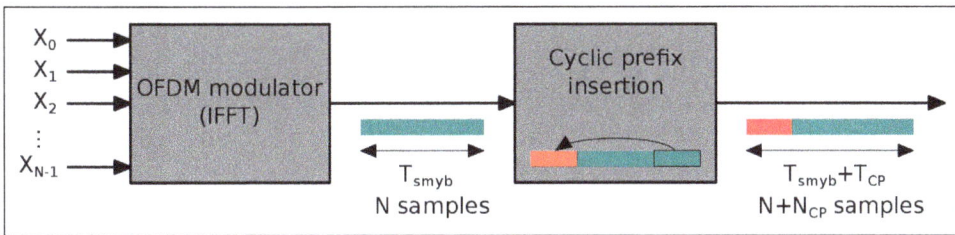

In some standards such as Ultrawideband, in the interest of transmitted power, cyclic prefix is skipped and nothing is sent during the guard interval. The receiver will then have to mimic the cyclic prefix functionality by copying the end part of the OFDM symbol and adding it to the beginning portion.

## Simplified Equalization

The effects of frequency-selective channel conditions, for example fading caused by multipath propagation, can be considered as constant (flat) over an OFDM sub-channel if the sub-channel is sufficiently narrow-banded (i.e., if the number of sub-channels is sufficiently large). This makes frequency domain equalization possible at the receiver, which is far simpler than the time-domain equalization used in conventional single-carrier modulation. In OFDM, the equalizer only has to multiply each detected subcarrier (each Fourier coefficient) in each OFDM symbol by a constant complex number, or a rarely changed value. On a fundamental level, simpler digital equalizers are better because they require fewer operations, which translates to fewer round-off errors in the equalizer. Those round-off errors can be viewed as numerical noise and are inevitable.

The OFDM equalization in the above numerical example would require one complex valued multiplication per subcarrier and symbol $N$=1000 complex multiplications per OFDM symbol; i.e., one million multiplications per second, at the receiver). The FFT algorithm requires $N \log_2 N$=10,000 [this is imprecise: over half of these complex multiplications are trivial, i.e. = to 1 and are not implemented in software or HW].

complex-valued multiplications per OFDM symbol (i.e., 10 million multiplications per second), at both the receiver and transmitter side. This should be compared with the corresponding one million symbols/second single-carrier modulation case mentioned in the example, where the equalization of 125 microseconds time-spreading using a FIR filter would require, in a naive implementation, 125 multiplications per symbol (i.e., 125 million multiplications per second). FFT techniques can be used to reduce the number of multiplications for an FIR filter-based time-domain equalizer to a number comparable with OFDM, at the cost of delay between reception and decoding which also becomes comparable with OFDM.

If differential modulation such as DPSK or DQPSK is applied to each subcarrier, equalization can be completely omitted, since these non-coherent schemes are insensitive to slowly changing amplitude and phase distortion.

In a sense, improvements in FIR equalization using FFTs or partial FFTs leads mathematically closer to OFDM, but the OFDM technique is easier to understand and implement, and the sub-channels can be independently adapted in other ways than varying equalization coefficients, such as switching between different QAM constellation patterns and error-correction schemes to match individual sub-channel noise and interference characteristics.

Some of the subcarriers in some of the OFDM symbols may carry pilot signals for measurement of the channel conditions (i.e., the equalizer gain and phase shift for each subcarrier). Pilot signals and training symbols (preambles) may also be used for time synchronization (to avoid intersymbol interference, ISI) and frequency synchronization (to avoid inter-carrier interference, ICI, caused by Doppler shift).

OFDM was initially used for wired and stationary wireless communications. However, with an increasing number of applications operating in highly mobile environments, the effect of dispersive fading caused by a combination of multi-path propagation and doppler shift is more significant. Over the last decade, research has been done on how to equalize OFDM transmission over doubly selective channels.

## Channel Coding and Interleaving

OFDM is invariably used in conjunction with channel coding (forward error correction), and almost always uses frequency and time interleaving.

Frequency (subcarrier) interleaving increases resistance to frequency-selective channel conditions such as fading. For example, when a part of the channel bandwidth fades, frequency interleaving ensures that the bit errors that would result from those subcarriers in the faded part of the bandwidth are spread out in the bit-stream rather than being concentrated. Similarly, time interleaving ensures that bits that are originally close together in the bit-stream are transmitted far apart in time, thus mitigating against severe fading as would happen when travelling at high speed.

However, time interleaving is of little benefit in slowly fading channels, such as for stationary reception, and frequency interleaving offers little to no benefit for narrowband channels that suffer from flat-fading (where the whole channel bandwidth fades at the same time).

The reason why interleaving is used on OFDM is to attempt to spread the errors out in the bit-stream that is presented to the error correction decoder, because when such decoders are presented with a high concentration of errors the decoder is unable to correct all the bit errors, and a burst of uncorrected errors occurs. A similar design of audio data encoding makes compact disc (CD) playback robust.

A classical type of error correction coding used with OFDM-based systems is convolutional coding, often concatenated with Reed-Solomon coding. Usually, additional interleaving in between the two layers of coding is implemented. The choice for Reed-Solomon coding as the outer error correction code is based on the observation that the Viterbi decoder used for inner convolutional decoding produces short error bursts when there is a high concentration of errors, and Reed-Solomon codes are inherently well suited to correcting bursts of errors.

Newer systems, however, usually now adopt near-optimal types of error correction codes that use the turbo decoding principle, where the decoder iterates towards the desired solution. Examples of such error correction coding types include turbo codes and LDPC codes, which perform close to the Shannon limit for the Additive White Gaussian Noise (AWGN) channel. Some systems that have implemented these codes have concatenated them with either Reed-Solomon (for example on the MediaFLO system) or BCH codes (on the DVB-S2 system) to improve upon an error floor inherent to these codes at high signal-to-noise ratios.

## Adaptive Transmission

The resilience to severe channel conditions can be further enhanced if information about the channel is sent over a return-channel. Based on this feedback information, adaptive modulation, channel coding and power allocation may be applied across all subcarriers, or individually to each subcarrier. In the latter case, if a particular range of frequencies suffers from interference or attenuation, the carriers within that range can be disabled or made to run slower by applying more robust modulation or error coding to those subcarriers.

The term discrete multitone modulation (DMT) denotes OFDM-based communication systems that adapt the transmission to the channel conditions individually for each subcarrier, by means of so-called bit-loading. Examples are ADSL and VDSL.

The upstream and downstream speeds can be varied by allocating either more or fewer carriers for each purpose. Some forms of rate-adaptive DSL use this feature in real time, so that the bitrate is adapted to the co-channel interference and bandwidth is allocated to whichever subscriber needs it most.

# Orthogonal Frequency-division Multiple Access

Orthogonal frequency-division multiple access (OFDMA) is a multi-user version of the popular orthogonal frequency-division multiplexing (OFDM) digital modulation scheme. Multiple access is achieved in OFDMA by assigning subsets of subcarriers to individual users. This allows simultaneous low-data-rate transmission from several users.

## Advantages and Disadvantages

Claimed advantages over OFDM with time-domain statistical multiplexing:

- Allows simultaneous low-data-rate transmission from several users.
- Pulsed carrier can be avoided.
- Lower maximal transmission power for low-data-rate users.
- Shorter delay and constant delay.
- Contention-based multiple access (collision avoidance) is simplified.
- Further improves OFDM robustness to fading and interference.
- Combat narrow-band interference.

## Claimed OFDMA Advantages

- Flexibility of deployment across various frequency bands with little needed modification to the air interface.
- Averaging interferences from neighbouring cells, by using different basic carrier permutations between users in different cells.
- Interferences within the cell are averaged by using allocation with cyclic permutations.
- Enables single-frequency network coverage, where coverage problem exists and gives excellent coverage.
- Offers frequency diversity by spreading the carriers all over the used spectrum.
- Allows per-channel or per-subchannel power.

## Recognised Disadvantages of OFDMA

- Higher sensitivity to frequency offsets and phase noise.
- Asynchronous data communication services such as web access are characterised by short communication bursts at high data rate. Few users in a base station cell are transferring data simultaneously at low constant data rate.

- The complex OFDM electronics, including the FFT algorithm and forward error correction, are constantly active independent of the data rate, which is inefficient from power-consumption point of view, while OFDM combined with data packet scheduling may allow FFT algorithm to hibernate during certain time intervals.

- The OFDM diversity gain and resistance to frequency-selective fading may partly be lost if very few sub-carriers are assigned to each user, and if the same carrier is used in every OFDM symbol. Adaptive sub-carrier assignment based on fast feedback information about the channel, or sub-carrier frequency hopping, is therefore desirable.

- Dealing with co-channel interference from nearby cells is more complex in OFDM than in CDMA. It would require dynamic channel allocation with advanced coordination among adjacent base stations.

- The fast channel feedback information and adaptive sub-carrier assignment is more complex than CDMA fast power control.

## Characteristics and Principles of Operation

Based on feedback information about the channel conditions, adaptive user-to-subcarrier assignment can be achieved. If the assignment is done sufficiently fast, this further improves the OFDM robustness to fast fading and narrow-band cochannel interference, and makes it possible to achieve even better system spectral efficiency.

Different numbers of sub-carriers can be assigned to different users, in view to support differentiated Quality of Service (QoS), i.e. to control the data rate and error probability individually for each user.

OFDMA can be seen as an alternative to combining OFDM with time-division multiple access (TDMA) or time-domain statistical multiplexing communication. Low-data-rate users can send continuously with low transmission power instead of using a "pulsed" high-power carrier. Constant delay, and shorter delay, can be achieved.

OFDMA can also be described as a combination of frequency-domain and time-domain multiple access, where the resources are partitioned in the time–frequency space, and slots are assigned along the OFDM symbol index, as well as OFDM sub-carrier index.

OFDMA is considered as highly suitable for broadband wireless networks, due to advantages including scalability and use of multiple antennas (MIMO)-friendliness, and ability to take advantage of channel frequency selectivity.

In spectrum sensing cognitive radio, OFDMA is a possible approach to filling free radio frequency bands adaptively. Timo A. Weiss and Friedrich K. Jondral of the University of Karlsruhe proposed a spectrum pooling system in which free bands sensed by nodes were immediately filled by OFDMA subbands.

## Usage

OFDMA is used in:

- The mobility mode of the IEEE 802.16 Wireless MAN standard, commonly referred to as WiMAX.

- The wireless LAN (WLAN) standard IEEE 802.11ax.

- The IEEE 802.20 mobile Wireless MAN standard, commonly referred to as MBWA.

- MoCA 2.0.

- The downlink of the 3GPP Long-Term Evolution (LTE) fourth-generation mobile broadband standard. The radio interface was formerly named *High Speed OFDM Packet Access* (HSOPA), now named Evolved UMTS Terrestrial Radio Access (E-UTRA).

- The downlink and the uplink of the 3GPP 5G New Radio (NR) fifth-generation mobile network standard. 5G NR is the successor to LTE.

- The Qualcomm Flarion Technologies Mobile Flash-OFDM.

- The now defunct Qualcomm/3GPP2 Ultra Mobile Broadband (UMB) project, intended as a successor of CDMA2000, but replaced by LTE.

OFDMA is also a candidate access method for the IEEE 802.22 *Wireless Regional Area Networks* (WRAN), a cognitive radio technology which uses white spaces in the television (TV) frequency spectrum, and the proposed access method for DECT-5G specification which aims to fulfill IMT-2020 requirements for high-throughput mobile broadband (eMMB) and ultra reliable low latency(URLLC) applications.

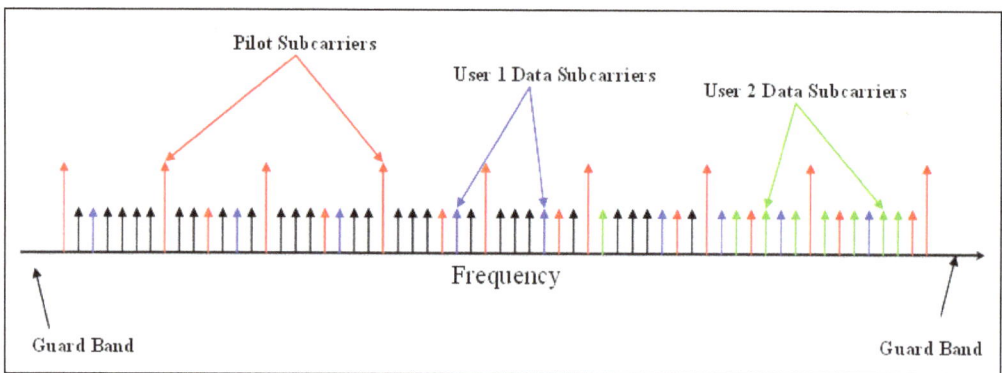

OFDMA Subcarriers.

## Space Diversity

In OFDM-based wide-area broadcasting, receivers can benefit from receiving signals from several spatially dispersed transmitters simultaneously, since transmitters will only destructively interfere with each other on a limited number of subcarriers, whereas in general they will actually reinforce coverage over a wide area. This is very beneficial in many countries, as it permits the operation of national single-frequency networks (SFN), where many transmitters send the same signal simultaneously over the same channel frequency.

FNs use the available spectrum more effectively than conventional multi-frequency broadcast networks (MFN), where program content is replicated on different carrier frequencies. SFNs also result in a diversity gain in receivers situated midway between the transmitters. The coverage area is increased and the outage probability decreased in comparison to an MFN, due to increased received signal strength averaged over all subcarriers.

Although the guard interval only contains redundant data, which means that it reduces the capacity, some OFDM-based systems, such as some of the broadcasting systems, deliberately use a long guard interval in order to allow the transmitters to be spaced farther apart in an SFN, and longer guard intervals allow larger SFN cell-sizes. A rule of thumb for the maximum distance between transmitters in an SFN is equal to the distance a signal travels during the guard interval — for instance, a guard interval of 200 microseconds would allow transmitters to be spaced 60 km apart.

A single frequency network is a form of transmitter macrodiversity. The concept can be further used in dynamic single-frequency networks (DSFN), where the SFN grouping is changed from timeslot to timeslot.

OFDM may be combined with other forms of space diversity, for example antenna arrays and MIMO channels. This is done in the IEEE 802.11 Wireless LAN standards.

## Linear Transmitter Power Amplifier

An OFDM signal exhibits a high peak-to-average power ratio (PAPR) because the independent phases of the subcarriers mean that they will often combine constructively. Handling this high PAPR requires:

- A high-resolution digital-to-analog converter (DAC) in the transmitter.

- A high-resolution analog-to-digital converter (ADC) in the receiver.

- A linear signal chain.

Any non-linearity in the signal chain will cause intermodulation distortion that:

- Raises the noise floor.

- May cause inter-carrier interference.

- Generates out-of-band spurious radiation.

The linearity requirement is demanding, especially for transmitter RF output circuitry where amplifiers are often designed to be non-linear in order to minimise power consumption. In practical OFDM systems a small amount of peak clipping is allowed to limit the PAPR in a judicious trade-off against the above consequences. However, the transmitter output filter which is required to reduce out-of-band spurs to legal levels has the effect of restoring peak levels that were clipped, so clipping is not an effective way to reduce PAPR.

Although the spectral efficiency of OFDM is attractive for both terrestrial and space communications, the high PAPR requirements have so far limited OFDM applications to terrestrial systems.

The crest factor CF (in dB) for an OFDM system with n uncorrelated subcarriers is:

$$CF = 10\log(n) + CF_c$$

where $CF_c$ is the crest factor (in dB) for each subcarrier. ($CF_c$ is 3.01 dB for the sine waves used for BPSK and QPSK modulation).

For example, the DVB-T signal in 2K mode is composed of 1705 subcarriers that are each QPSK-modulated, giving a crest factor of 35.32 dB.

Many crest factor reduction techniques have been developed.

The dynamic range required for an FM receiver is 120 dB while DAB only require about 90 dB. As a comparison, each extra bit per sample increases the dynamic range with 6 dB.

## Efficiency Comparison between Single Carrier and Multicarrier

The performance of any communication system can be measured in terms of its power efficiency and bandwidth efficiency. The power efficiency describes the ability of communication system to preserve bit error rate (BER) of the transmitted signal at low power levels. Bandwidth efficiency reflects how efficiently the allocated bandwidth is used and is defined as the throughput data rate per hertz in a given bandwidth. If the large number of subcarriers are used, the bandwidth efficiency of multicarrier system such as OFDM with using optical fiber channel is defined as:

$$\eta = 2 \cdot \frac{R_s}{B_{OFDM}}$$

where $R_s$ is the symbol rate in giga-symbols per second (Gsps), $B_{OFDM}$ is the bandwidth of OFDM signal, and the factor of 2 is due to the two polarization states in the fiber.

There is saving of bandwidth by using multicarrier modulation with orthogonal frequency division multiplexing. So the bandwidth for multicarrier system is less in comparison with single carrier system and hence bandwidth efficiency of multicarrier system is larger than single carrier system.

| S.no. | Transmission type | M in M-QAM | No. of sub-carriers | Bit rate | Fiber length | Power at the receiver (at BER of $10^{-9}$) | Bandwidth efficiency |
|---|---|---|---|---|---|---|---|
| 1. | Single carrier | 64 | 1 | 10 Gbit/s | 20 km | $-37.3$ dBm | 6.0000 |
| 2. | Multicarrier | 64 | 128 | 10 Gbit/s | 20 km | $-36.3$ dBm | 10.6022 |

There is only 1 dBm increase in receiver power, but we get 76.7% improvement in bandwidth efficiency with using multicarrier transmission technique.

## Idealized System Model

## Transmitter

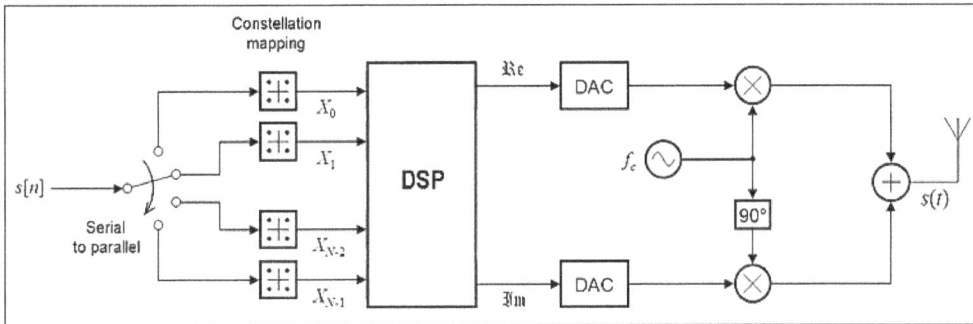

An OFDM carrier signal is the sum of a number of orthogonal subcarriers, with baseband data on each subcarrier being independently modulated commonly using some type of quadrature amplitude modulation (QAM) or phase-shift keying (PSK). This composite baseband signal is typically used to modulate a main RF carrier.

$s[n]$ is a serial stream of binary digits. By inverse multiplexing, these are first demultiplexed into $N$ parallel streams, and each one mapped to a (possibly complex) symbol stream using some modulation constellation (QAM, PSK, etc.). Note that the constellations may be different, so some streams may carry a higher bit-rate than others.

An inverse FFT is computed on each set of symbols, giving a set of complex time-domain samples. These samples are then quadrature-mixed to passband in the standard way. The real and imaginary components are first converted to the analogue domain using digital-to-analogue converters (DACs); the analogue signals are then used to modulate cosine and sine waves at the carrier frequency, $f_c$, respectively. These signals are then summed to give the transmission signal, $s(t)$.

## Receiver

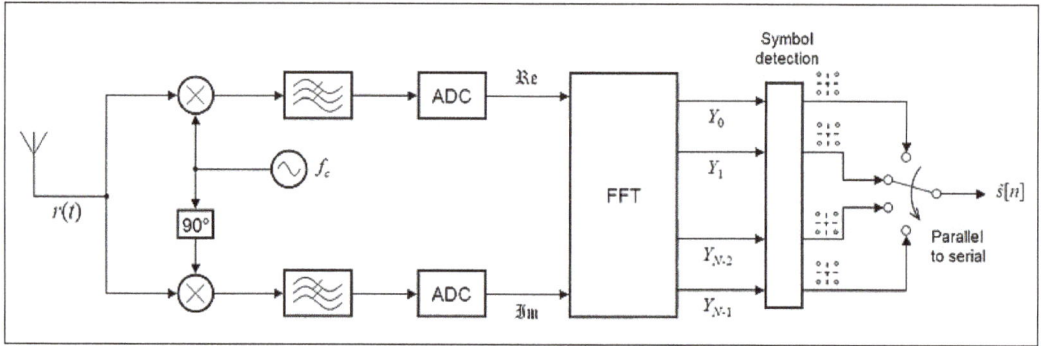

The receiver picks up the signal $r(t)$, which is then quadrature-mixed down to baseband using cosine and sine waves at the carrier frequency. This also creates signals centered on $2f_c$, so low-pass filters are used to reject these. The baseband signals are then sampled and digitised using analog-to-digital converters (ADCs), and a forward FFT is used to convert back to the frequency domain.

This returns $N$ parallel streams, each of which is converted to a binary stream using an appropriate symbol detector. These streams are then re-combined into a serial stream, $\hat{s}[n]$, which is an estimate of the original binary stream at the transmitter.

## Mathematical Description

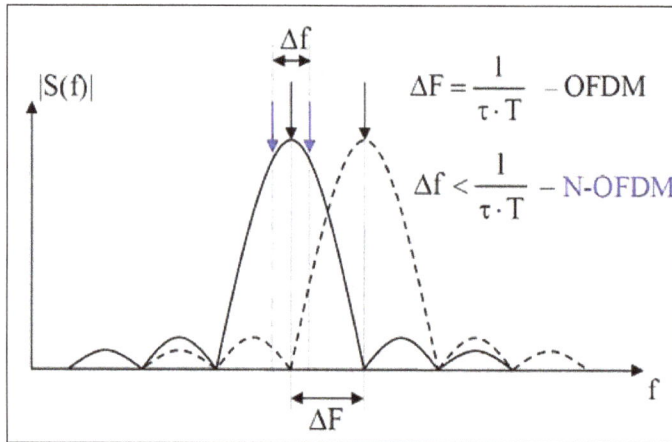

Subcarriers system of OFDM signals after FFT.

If $N$ subcarriers are used, and each subcarrier is modulated using $M$ alternative symbols, the OFDM symbol alphabet consists of $M^N$ combined symbols.

The low-pass equivalent OFDM signal is expressed as:

$$v(t) = \sum_{k=0}^{N-1} X_k e^{j2\pi kt/T}, \quad 0 \le t < T,$$

where $\{X_k\}$ are the data symbols, $N$ is the number of subcarriers, and $T$ is the OFDM symbol time. The subcarrier spacing of $\dfrac{1}{T}$ makes them orthogonal over each symbol period; this property is expressed as:

$$\frac{1}{T}\int_0^T \left(e^{j2\pi k_1 t/T}\right)^* \left(e^{j2\pi k_2 t/T}\right)dt$$

$$=\frac{1}{T}\int_0^T e^{j2\pi(k_2-k_1)t/T}dt = \delta_{k_1 k_2}$$

where $(\cdot)^*$ denotes the complex conjugate operator and $\delta$ is the Kronecker delta.

To avoid intersymbol interference in multipath fading channels, a guard interval of length $T_g$ is inserted prior to the OFDM block. During this interval, a *cyclic prefix* is transmitted such that the signal in the interval $-T_g \leq t < 0$ equals the signal in the interval $(T-T_g) \leq t < T$. The OFDM signal with cyclic prefix is thus:

$$v(t) = \sum_{k=0}^{N-1} X_k e^{j2\pi kt/T}, \quad -T_g \leq t < T$$

The low-pass signal above can be either real or complex-valued. Real-valued low-pass equivalent signals are typically transmitted at baseband—wireline applications such as DSL use this approach. For wireless applications, the low-pass signal is typically complex-valued; in which case, the transmitted signal is up-converted to a carrier frequency $f_c$. In general, the transmitted signal can be represented as:

$$s(t) = \Re\left\{v(t)e^{j2\pi f_c t}\right\}$$

$$= \sum_{k=0}^{N-1} |X_k|\cos\left(2\pi[f_c + k/T]t + \arg[X_k]\right)$$

## Usage

OFDM is used in:

- Digital Radio Mondiale DRM,

- Digital Audio Broadcasting (DAB),

- Digital television DVB-T/T2 (terrestrial), DVB-H (handheld), DMB-T/H, DVB-C2 (cable),

- Wireless LAN IEEE 802.11a, IEEE 802.11g, IEEE 802.11n, IEEE 802.11ac, and IEEE 802.11ad,

- WiMAX,

- Li-Fi,

- ADSL (G.dmt/ITU G.992.1),

- The LTE and LTE Advanced 4G mobile phone standards,

- Modern narrow and broadband power line communications.

## ADSL

OFDM is used in ADSL connections that follow the ANSI T1.413 and G.dmt (ITU G.992.1) standards, where it is called *discrete multitone modulation* (DMT). DSL achieves high-speed data connections on existing copper wires. OFDM is also used in the successor standards ADSL2, ADSL2+, VDSL, VDSL2, and G.fast. ADSL2 uses variable subcarrier modulation, ranging from BPSK to 32768QAM (in ADSL terminology this is referred to as bit-loading, or bit per tone, 1 to 15 bits per subcarrier).

Long copper wires suffer from attenuation at high frequencies. The fact that OFDM can cope with this frequency selective attenuation and with narrow-band interference are the main reasons it is frequently used in applications such as ADSL modems.

## Powerline Technology

OFDM is used by many powerline devices to extend digital connections through power wiring. Adaptive modulation is particularly important with such a noisy channel as electrical wiring. Some medium speed smart metering modems, "Prime" and "G3" use OFDM at modest frequencies (30–100 kHz) with modest numbers of channels (several hundred) in order to overcome the intersymbol interference in the power line environment. The IEEE 1901 standards include two incompatible physical layers that both use OFDM. The ITU-T G.hn standard, which provides high-speed local area networking over existing home wiring (power lines, phone lines and coaxial cables) is based on a PHY layer that specifies OFDM with adaptive modulation and a Low-Density Parity-Check (LDPC) FEC code.

## Wireless Local Area Networks (LAN) and Metropolitan Area Networks (MAN)

OFDM is extensively used in wireless LAN and MAN applications, including IEEE 802.11a/g/n and WiMAX.

IEEE 802.11a/g/n, operating in the 2.4 and 5 GHz bands, specifies per-stream airside data rates ranging from 6 to 54 Mbit/s. If both devices can use "HT mode" (added with 802.11n), the top 20 MHz per-stream rate is increased to 72.2 Mbit/s, with the option of data rates between 13.5 and 150 Mbit/s using a 40 MHz channel. Four different modulation schemes are used: BPSK, QPSK, 16-QAM, and 64-QAM, along with a set of error correcting rates (1/2–5/6). The multitude of choices allows the system to adapt the optimum data rate for the current signal conditions.

## Wireless Personal Area Networks (PAN)

OFDM is also now being used in the WiMedia/Ecma-368 standard for high-speed wireless personal area networks in the 3.1–10.6 GHz ultrawideband spectrum.

## Terrestrial Digital Radio and Television Broadcasting

Much of Europe and Asia has adopted OFDM for terrestrial broadcasting of digital television (DVB-T, DVB-H and T-DMB) and radio (EUREKA 147 DAB, Digital Radio Mondiale, HD Radio and T-DMB).

## DVB-T

By Directive of the European Commission, all television services transmitted to viewers in the European Community must use a transmission system that has been standardized by a recognized European standardization body, and such a standard has been developed and codified by the DVB Project, *Digital Video Broadcasting (DVB); Framing structure, channel coding and modulation for digital terrestrial television.* Customarily referred to as DVB-T, the standard calls for the exclusive use of COFDM for modulation. DVB-T is now widely used in Europe and elsewhere for terrestrial digital TV.

## SDARS

The ground segments of the Digital Audio Radio Service (SDARS) systems used by XM Satellite Radio and Sirius Satellite Radio are transmitted using Coded OFDM (COFDM). The word "coded" comes from the use of forward error correction (FEC).

## COFDM vs VSB

The question of the relative technical merits of COFDM versus 8VSB for terrestrial digital television has been a subject of some controversy, especially between European and North American technologists and regulators. The United States has rejected several proposals to adopt the COFDM-based DVB-T system for its digital television services, and has instead opted for 8VSB (vestigial sideband modulation) operation.

One of the major benefits provided by COFDM is in rendering radio broadcasts relatively immune to multipath distortion and signal fading due to atmospheric conditions or passing aircraft. Proponents of COFDM argue it resists multipath far better than 8VSB. Early 8VSB DTV (digital television) receivers often had difficulty receiving a signal. Also, COFDM allows single-frequency networks, which is not possible with 8VSB.

However, newer 8VSB receivers are far better at dealing with multipath, hence the difference in performance may diminish with advances in equalizer design.

## Digital Radio

COFDM is also used for other radio standards, for Digital Audio Broadcasting (DAB), the standard for digital audio broadcasting at VHF frequencies, for Digital Radio Mondiale (DRM), the standard for digital broadcasting at shortwave and medium wave frequencies (below 30 MHz) and for DRM+ a more recently introduced standard for digital audio broadcasting at VHF frequencies. (30 to 174 MHz).

The USA again uses an alternate standard, a proprietary system developed by iBiquity dubbed *HD Radio*. However, it uses COFDM as the underlying broadcast technology to add digital audio to AM (medium wave) and FM broadcasts.

Both Digital Radio Mondiale and HD Radio are classified as in-band on-channel systems, unlike Eureka 147 (DAB: Digital Audio Broadcasting) which uses separate VHF or UHF frequency bands instead.

## BST-OFDM used in ISDB

The *band-segmented transmission orthogonal frequency division multiplexing (BST-OFDM)* system proposed for Japan (in the ISDB-T, ISDB-TSB, and ISDB-C broadcasting systems) improves upon COFDM by exploiting the fact that some OFDM carriers may be modulated differently from others within the same multiplex. Some forms of COFDM already offer this kind of hierarchical modulation, though BST-OFDM is intended to make it more flexible. The 6 MHz television channel may therefore be "segmented", with different segments being modulated differently and used for different services.

It is possible, for example, to send an audio service on a segment that includes a segment composed of a number of carriers, a data service on another segment and a television service on yet another segment—all within the same 6 MHz television channel. Furthermore, these may be modulated with different parameters so that, for example, the audio and data services could be optimized for mobile reception, while the television service is optimized for stationary reception in a high-multipath environment.

## Ultra-wideband

Ultra-wideband (UWB) wireless personal area network technology may also use OFDM, such as in Multiband OFDM (MB-OFDM). This UWB specification is advocated by the WiMedia Alliance (formerly by both the Multiband OFDM Alliance [MBOA] and the WiMedia Alliance, but the two have now merged), and is one of the competing UWB radio interfaces.

## FLASH-OFDM

*Fast low-latency access with seamless handoff orthogonal frequency division multiplexing* (Flash-OFDM), also referred to as F-OFDM, was based on OFDM and also specified higher protocol layers. It was developed by Flarion, and purchased by Qualcomm in January 2006. Flash-OFDM was marketed as a packet-switched cellular bearer, to compete with GSM and 3G networks. As an example, 450 MHz frequency bands previously used by NMT-450 and C-Net C450 (both 1G analogue networks, now mostly decommissioned) in Europe are being licensed to Flash-OFDM operators.

Slovak Telekom in Slovakia offers Flash-OFDM connections with a maximum downstream speed of 5.3 Mbit/s, and a maximum upstream speed of 1.8 Mbit/s, with a coverage of over 70 percent of Slovak population. The Flash-OFDM network was switched off in the majority of Slovakia on 30 September 2015.

T-Mobile Germany used Flash-OFDM to backhaul Wi-Fi HotSpots on the Deutsche Bahn's ICE high speed trains between 2005 and 2015, until switching over to UMTS and LTE.

American wireless carrier Nextel Communications field tested wireless broadband network technologies including Flash-OFDM in 2005. Sprint purchased the carrier in 2006 and decided to deploy the mobile version of WiMAX, which is based on Scalable Orthogonal Frequency Division Multiple Access (SOFDMA) technology.

Citizens Telephone Cooperative launched a mobile broadband service based on Flash-OFDM technology to subscribers in parts of Virginia in March 2006. The maximum speed available was 1.5 Mbit/s. The service was discontinued on April 30, 2009.

## Wavelet-OFDM

OFDM has become an interesting technique for power line communications (PLC). In this area of research, a wavelet transform is introduced to replace the DFT as the method of creating orthogonal frequencies. This is due to the advantages wavelets offer, which are particularly useful on noisy power lines.

Instead of using an IDFT to create the sender signal, the wavelet OFDM uses a synthesis bank consisting of a $N$-band transmultiplexer followed by the transform function:

$$F_n(z) = \sum_{k=0}^{L-1} f_n(k)z^{-k}, \quad 0 \leq n < N$$

On the receiver side, an analysis bank is used to demodulate the signal again. This bank contains an inverse transform:

$$G_n(z) = \sum_{k=0}^{L-1} g_n(k)z^{-k}, \quad 0 \leq n < N$$

followed by another $N$-band transmultiplexer. The relationship between both transform functions is:

$$f_n(k) = g_n(L-1-k)$$
$$F_n(z) = z^{-(L-1)}G_n *(z-1)$$

An example of W-OFDM uses the Perfect Reconstruction Cosine Modulated Filter Bank (PR-CMFB) and Extended Lapped Transform (ELT) is used for the wavelet TF. Thus, $f_n(k)$ and $g_n(k)$ are given as:

$$f_n(k) = 2p_0(k)\cos\left[\frac{\pi}{N}\left(n+\frac{1}{2}\right)\left(k-\frac{L-1}{2}\right) - (-1)^n\frac{\pi}{4}\right]$$

$$g_n(k) = 2p_0(k)\cos\left[\frac{\pi}{N}\left(n+\frac{1}{2}\right)\left(k-\frac{L-1}{2}\right) + (-1)^n\frac{\pi}{4}\right]$$

$$P_0(z) = \sum_{k=0}^{N-1} z^{-k}Y_k\left(z^{2N}\right)$$

These two functions are their respective inverses, and can be used to modulate and demodulate a given input sequence. Just as in the case of DFT, the wavelet transform creates orthogonal waves with, $f_0, f_1, ..., f_{N-1}$ The orthogonality ensures that they do not interfere with each other and can be sent simultaneously. At the receiver, $g_0, g_1, ..., g_{N-1}$ are used to reconstruct the data sequence once more.

## Advantages over Standard OFDM

W-OFDM is an evolution of the standard OFDM, with certain advantages. Mainly, the sidelobe levels of W-OFDM are lower. This results in less ICI, as well as greater robustness to narrowband interference. These two properties are especially useful in PLC, where most of the lines aren't shielded against EM-noise, which creates noisy channels and noise spikes.

A comparison between the two modulation techniques also reveals that the complexity of both algorithms remains approximately the same.

# MULTICARRIER MODULATION

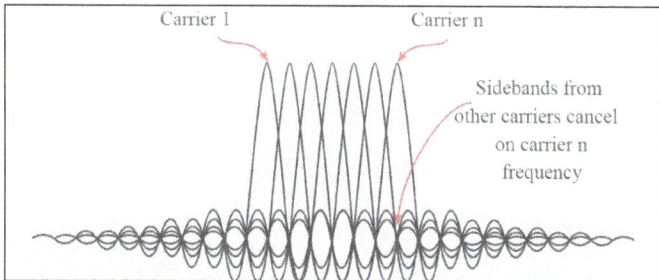

One form of multicarrier modulation is OFDM.

Multicarrier modulation, MCM is a technique for transmitting data by sending the data over multiple carriers which are normally close spaced.

Multicarrier modulation has several advantages including resilience to interference, resilience to narrow band fading and multipath effects.

As a result, multicarrier modulation techniques are widely used for data transmission as it is able to provide an effective signal waveform which is spectrally efficient and resilient to the real world environment.

## Multicarrier Modulation Basics

Multicarrier modulation operates by dividing the data stream to be transmitted into a number of lower data rate data streams. Each of the lower data rate streams is then used to modulate an individual carrier.

When the overall transmission is received, the receiver has to then re-assembles the overall data stream from those received on the individual carriers.

It is possible to use a variety of different techniques for multicarrier transmissions. Each form of MCM has its own advantages and can be sued in different applications.

## Multicarrier Modulation Systems

There are many forms of multicarrier modulation techniques that are in use of being investigated for future use. Some of the more widely known schemes are summarised below:

- Orthogonal frequency division multiplexing, OFDM: OFDM is possibly the most widely used form of multicarrier modulation. It uses multiple closely spaced carriers and as a result of their orthogonality, mutual interference between them is avoided.

- Generalised Frequency Division Multiplexing, GFDM: GFDM is a multicarrier modulation scheme that uses closed spaced non-orthogonal carriers and provides flexible pulse shaping. It is therefore attractive for various applications such as machine to machine communications.

- Filter Bank Multi Carrier, FBMC: FBMC is a form of multicarrier modulation scheme that uses a specialised pulse shaping filter known as an isotropic orthogonal transform algorithm, IOTA within the digital signal processing for the system. This scheme provides good time and frequency localisation properties and this ensures that inter-symbol interference and inter-carrier interference are avoided without the use of cyclic prefix required for OFDM based systems.

- SEFDM: Spectrally efficient frequency division multiplex uses multiple carriers in the same way as OFDM, but they are spaced closer than OFDM. However it is stil possible to recover the data, although with a slight power penalty.

The various forms of multicarrier modulation each have their own characteristics and advantages. This means that they are applicable in different circumstances, providing improvements in certain areas according to the type of multicarrier modulation used.

# RESOURCE ALLOCATION IN OFDM-BASED WIMAX

The worldwide interoperability for microwave access (WiMAX) extends the transmission rate and range of wireless communications beyond the limits of existing technologies while allowing for heterogeneous traffic transmissions. To achieve all these goals, qualified protocols for WiMAX should effectively utilize the spectrum and overcome the deficits of wireless channel while maintaining a satisfactory level of heterogeneous services for users. WiMAX supports air interfaces based on orthogonal frequency division multiplexing (OFDM) which is a robust and flexible technique for transmissions and resource allocations, respectively, over wireless channel.

WiMAX is expected to provide high data-rate services over a service area as large as a metropolitan area network. Broad spectrum and large coverage area usually cause severe interference and multipath transmissions, unless an appropriate design takes effect. WiMAX deploys multicarrier transmission based on the OFDM and multiple access based on the orthogonal frequency division multiple access (OFDMA). OFDM mitigates noise, multipath, and interference effects which are the main challenges of wireless communications. OFDMA is very flexible in allocating resources which are very critical for wireless networks.

The main scarce resources in wireless networks are spectrum and power. In spite of the large frequency band of the broadband networks, network designers allocate it very efficiently to serve as many satisfied users as possible while maintaining reasonable level of revenue for service providers. Mobile and portable devices are required to consume minimal power to extend their battery lifetime. Furthermore, fixed equipment and base stations (BSs) are expected to consume as low power as possible due to the health and global energy concerns. The resource constraints of the wireless medium become more critical when resource demanding applications are dealt with. According to the IEEE 802.16 standard, the coexistence of real-time and non-real-time traffic in WiMAX is promising. Therefore, existing resource allocation schemes that support one type of the traffics fails and the development of new schemes that simultaneously satisfy diverse quality-of-service (QoS) demands of the heterogeneous traffic and fairly manage resources becomes necessary.

## OFDM-based Wimax

The channel impairments significantly degrade the performance of a broadband wireless communication network.Multicarrier transmission is selected as a promisingtechnique for future communication due to its robustness against the frequency selectivity of broadband communications.

## Radio Channel

The wireless propagation channel constrains the information communication capacity between a transmitter and a receiver. The design of a wireless communication system's coding, modulation signal-processing algorithms, and multiple access scheme is predicated on the channel model. The wireless channel is also generally time-varying, space-varying, frequency-varying, polarizationvarying, dependent on the particular environment and the transmitter and receiver's location. Each type of variation presents randomness and unpredictability. Nevertheless, the communication channel modeling has been well established to characterize the channel by various time and frequency metrics.

A channel's impulse response to $\delta(\tau)$, Dirac impulse function transmitted at the moment $\tau$, looks like a series of impulses, because of the multipath reflections, represented by a time-variant function:

$$h(\tau,t) = \sum_{p=0}^{N_p-1} a_p(t) e^{j\left(2\pi f\,D,p'+\varphi p\right)} \, \delta\left(\tau - \tau_p(t)\right).$$

$a_p$, $f_{D,p}$, $\varphi_p$, and $\tau_p$, respectively refer to the $p$th multipath's complex-valued arrival amplitude, Doppler frequency, phase, and arrival excess delay, i.e., the delay measured with respect to the arrival of the first multipath component. $N_p$ symbolizes the number of multipaths whose amplitudes exceed the detection threshold. In practice, the number of multipath components that can be distinguished is very large. Therefore, only those multipaths that are temporally resolvable, i.e., their difference in arrival time to the receiver is greater than the inverse of the input signal bandwidth, are considered in detection.

The multipath propagation mechanisms (reflection, diffraction, and scattering) result in delay dispersion, which corresponds to frequency selectivity in the spectral domain. Each multipath power and delay is given by the power delay profile (PDP) denoted by $\left|h(\tau)\right|^2$, where h($\tau$) denotes the temporally stationary discrete-time channel's impulse response:

$$h(\tau) = \sum_{p=0}^{N_p-1} a_p \delta\left(\tau - \tau_p\right)$$

The channel's PDP is a mathematical function whose complete characterization necessitates a long signal. However, the scalar metrics mean delay:

$$\bar{\tau} = \frac{\int_0^\infty |h(t)|^2 t\,dt}{\int_0^\infty |h(t)|^2 t\,dt} = \frac{\sum_{p=0}^{Np-1} |(a_p)|^2 \tau_p}{\sum_{p=0}^{Np-1} |(a_p)|^2}$$

and RMS delay spread:

$$\tau_{\mathrm{RMS}} = \frac{\int_0^\infty \left|h(t)(t-\bar{\tau})^2\right| dt}{\int_0^\infty |h(t)|^2\, dt} = \frac{\sum_{p=0}^{Np-1} \left|a_p\left(\tau_p - \bar{\tau}\right)\right|^2}{\sum_{p=0}^{Np-1} |(a_p)|^2}$$

characterize the PDP temporal dispersion. Figure shows the graphical representations of the mean delay and RMS delay spread. When the channel delay dispersion is greater than the signal reciprocal bandwidth, i.e., the symbol duration $T_s \ll \tau_{\mathrm{RMS}}$, the transmitted train of symbols overlaps at the receiver. This phenomenon is known as inter-symbol interference (ISI) which is illustrated in figure.

Whereas the preceding metrics are in the relative-delay domain, channel fading may also be characterized in the spectral domain. Specifically, the coherence bandwidth, $B_c$, offers an alternative metric to the RMS-delay-spread, $\tau_{\mathrm{RMS}}$, to measure the channel's delay dispersion. The channel impulse response's Fourier transform gives the time-variant channel transfer function.

$$H(f,t) = \sum_{p=0}^{N_p-1} a_p(t)e^{j(2\pi(f\,D,pt-f\,\tau_p(t))+\varphi p)}$$

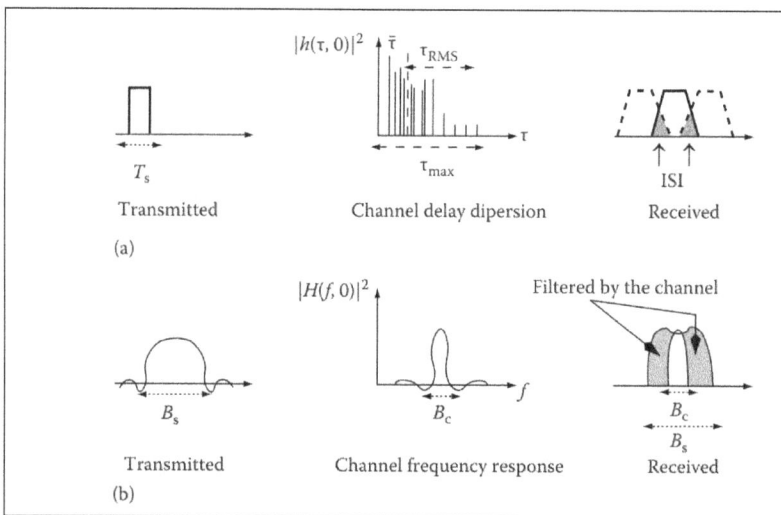

Wireless channel effect: (a) delay dispersion (b) frequency selectivity.

We define the channel frequency response's autocorrelation function as:

$$R(\Delta f) = E\{H(f, 0)H^*(f - \Delta f, 0)\}$$

where $(\cdot)*$ denotes the complex conjugate. The coherence bandwidth $B_c$ measures the spectral width of $|R(\Delta f)|$ over which the channel is considered frequency flat. Note that the frequency selectivity is relative to the transmitted signal bandwidth. In particular, if the channel's $B_c$ is less than the transmitted signal bandwidth, the channel distorts the received signal at selected frequencies as shown in figure. On the other hand, the channel does not affect the received signal, if its $B_c$ is greater than the transmitted signal bandwidth.

Independently from delay dispersion and frequency selectivity, user mobility causes frequency dispersion which in turn results in channel time selectivity. The time correlation function.

$$R(\Delta t) = E\{H(0, t)H^*(0, t - \Delta t)\}$$

quantifies the time-varying nature of the channel. From $R(\Delta t)$, the channel coherence time $T_c$ can be obtained, and it is defined as the time duration over which the channel is essentially flat. $R(\Delta t)$ Fourier transform is the channel Doppler power spectrum that its correlation width is the Doppler spread $B_d$. If the channel impulse response changes rapidly within the symbol duration, i.e., $T_c < T_s$, the transmitted signal undergoes fast fading which leads to signal distortion, or the Doppler spread $B_d$ is greater than the transmitted signal bandwidth. This effect induces frequency offset and possibly inter-carrier interference (ICI) in dense spectrums. In summary,

- Delay dispersion results in frequency selective fading that alters the received signal waveform and hence causes performance degradation. The channel effect can be avoided by transforming the broadband signal into parallel narrowband signals with bandwidth smaller than the channel's $B_c$.

- Frequency dispersion smears the signal spectrum in the frequency domain. It causes time selectivity that varies the signal at a rate higher than the rate at which the channel can be accurately estimated.

Frequency division multiplexing (FDM) appeared in 1950s; however, its implementation required multiple analog radio frequency (RF) modules in each transceiver that made FDM impractical . Recently, the implementation of IFFT/FFT and FDM ability in mitigating the channels ISI brought FDM back under light. While FDM major advantage is eliminating the ISI effect, it does not eliminate the ICI that rises due to closely packed multicarriers. Alternatively, data symbols can be modulated on orthogonal multiple carriers to reduce ICI, which is termed OFDM.

The high data-rate stream is partitioned into $B$ data blocks of $N_{sc}$ length at the transmitter. The symbols are serial-to-parallel converted which increases the source symbol duration $T_s$ to:

$$T_S' = N_{sc}T_s.$$

As the symbol duration increases, the ISI effect significantly decreases. Thus, the need for an equalizer at the receiver is eliminated, which reduces the complexity of the receiver. After serial-to-parallel conversion, block b represents the OFDM symbol which consists of $N_{sc}$ complex data symbols denoted by {sn, b}, n = 0, ... , $N_{sc}$ − 1, b = 1, ... , B. For simplicity, the $b$th index is dropped and we only refer to the $s_n$, $n = 0,... , N_{sc} − 1$ sequence in each block, i.e., the OFDM symbol, unless it is necessary. Figure depicts a typical multiuser OFDM/OFDMA transmitter and receiver block diagram. OFDM is implemented by applying inverse discrete Fourier transform (IDFT) to the data sequence $s_n$ which gives the following samples $xv$, $v = 0, ..., N_{sc} − 1$, of each OFDM symbol:

$$x_v = \frac{1}{2}\sum_{n=0}^{N_{sc}} S_n e^{j\frac{\partial nv}{N_{sc}}} \quad v=0,..,N_{SC} -1.$$

Whereas serial-to-parallel conversion only reduces the ISI effect, cyclic extension of the symbol by inserting a guard interval $T_g$ that is longer than the maximum channel dispersion time $\tau_{max}$ eliminates the residual ISI effect. The guard interval is a copy of $T_S'$ last $L_g = \left\lceil \dfrac{\tau_{max} N_{sc}}{T_S'} \right\rceil$ samples (practically, the symbol source is continuous and guard insertion is archived by adjusting the starting phase and making the symbol period longer). After cyclic extension of the OFDM symbol, the time domain sampled sequence becomes.

$$x_v' = \frac{1}{N_{sc}} \sum_{n=0}^{N_{sc}-1} S_n e^{j\frac{2\partial nv}{N_{sc}}} \quad v = -L_g, \ ... \ , N_{sc} - 1$$

Then, the sequence $x'$ is passed through digital-to-analog converter, and its output is transmitted through the wireless channel. Figure depicts the time and frequency representation of an OFDM frame.

By implementing the OFDM multicarrier modulation, the continuous channel transfer function is sampled in time at the OFDM symbol rate $1/T_S''$ and in frequency at spacing Fs. The discrete channel transfer function adapted to multicarrier signals is given by:

$$H_{n,i} = H\left(nF_s, iT_s''\right)$$

$$\sum_{p=0}^{N_p-1} \alpha_p(t) e^{j\left(2\pi\left(f_D, piT_s''-nF s\tau_p(t)+\varphi p\right)\right)}$$

$$a_{n,i} e^{j\varphi n,i}$$

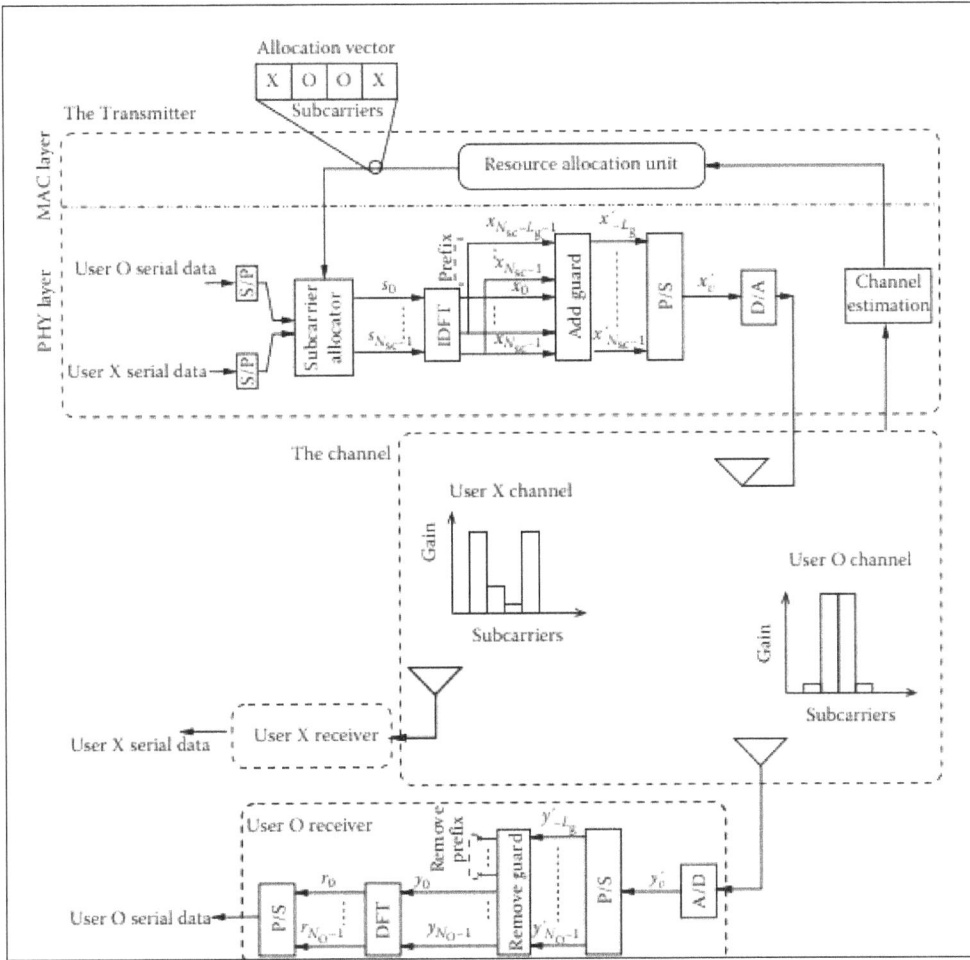

OFDMA transmitter and receiver's PHY and MAC structure.

A transmitted symbol on subcarrier $j$ of the OFDM symbol $b$ is multiplied by the resulting fading amplitude $a_{n,b}$ and rotated by random phase $\varphi_{n,i}$. The subcarrier gains can be represented by the following $N_{sc} \times N_{sc}$ channel matrix for the OFDM symbol $i$,

$$H = \left[ H_{0,0} \ H_{1,1} \ \dots \ H_{Nsc-1,N_{sc}-1} \right] \times I,$$

where $I$ is the identity matrix. The OFDM symbol index $i$ has been dropped for simplicity. Let $h_v$ be the sampled $L$ sequence of the channel impulse response $h(\tau, t)$ given in Equation 6.1 at a particular time instant $t$, i.e., $h_v = h\left( lT_s'', bT_s'' \right)$, $l = 0,\dots, L = \left( \tau_{max} / T_s'' \right)$, and represented by the vector h. Then, the matrix H diagonal elements are the discrete Fourier transform (DFT) of channel discrete impulse response.

After analog-to-digital conversion at the receiver, the received sampled sequence $y_v'$, $v = -L_g, \ \dots \ , N_{sc} - 1$ contains ISI in the first $L_g$ samples that are discarded. The remaining sequence $0, \ \dots \ , N_{sc} - 1$ is demodulated by the DFT. The DFT demodulated multicarrier sequence $r_n$, $n = 0, \ldots, N_{sc} - 1$, consists of $N_{sc}$ complex valued symbols:

$$r_n = \sum_{n=0}^{N_{sc}-1} y_v e^{-j\frac{2\delta nv}{N_{sc}}}, n = -0, \ldots, N_{SC} - 1.$$

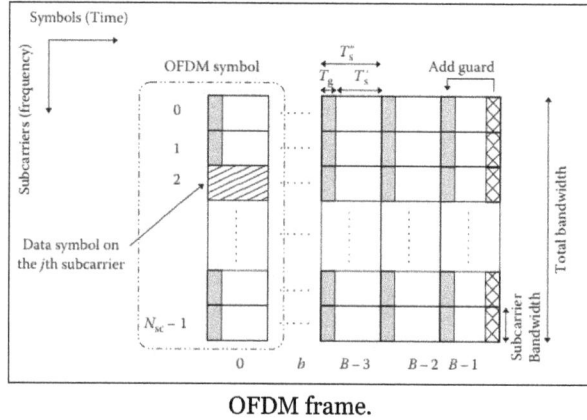

OFDM frame.

Since $x_v$ and $y_v$ are the sampled sequences of the transmitted and received signals, respectively, the vector representation of the received data symbols is given by:

$$r = \left[ r_0 \, r_1 \ldots r_{N_{SC}-1} \right]^T$$
$$= W^H y.$$

The operator $(\cdot)^H$ denotes the matrix Hermitian, y is defined as $\left[ y_0 \, y_1 \ldots y N_{sc}-1 \right]^T$, and W is the $N_{sc} \times N_{sc}$ IDFT matrix. The received signal in the frequency domain cannot be represented by the multiplication of the transmitted and the channel frequency domain representation because $y_v$ is related to $x_v$ and $h_v$ by linear convolution rather than circular convolution. However, the cyclic convolution is created for OFDM by appending the $L_g$ samples $\left( xN_{sc} - L_g - 1, xN_{sc} - L_g, \ldots, xN_{sc} - 1 \right)$ to the sequence $x_v$. This circular convolution of the two periodic sequences is transformed into the product of their DFTs.

$$r_n = H_n s_n \quad n = 0, \ldots, N_{sc} - 1$$

which can be alternatively written as:

$$r = Hs$$

r and s, respectively, are the received symbols and transmitted symbols matrix representation. Ignoring the additive noise effect and substituting equation $r = \left[ r_0 \, r_1 \ldots r_{N_{SC}-1} \right]^T = W^H y$ in the left-hand-side of equation $r = Hs$ result in:

$$W^H y^T = Hs$$
$$H^{-1} W^H y^T = s,$$

where $H^{-1}$ is the matrix inverse (the inverse of a diagonal matrix is a matrix with diagonal elements $\frac{1}{H_{n,n}}$ $n = 0,..., N_{sc} - 1$ ). Therefore, based on the availability of the channel estimation matrix H and by the implementation of DFT, the transmitted symbols can simply be recovered.

## Transmission Rate

Multicarrier modulation transforms a wide band channel experiencing selective fading onto multiple bands that experience flat fading. The flat fading channel is assumed to be static over the OFDM symbol duration. In addition, a perfect CSI is assumed to be available at the transmitter. Under these assumptions, the normalized transmission rate (bits/seconds/hertz) on the jth subcarrier is given by:

$$r_j = \log_2\left(1 + p_j \frac{|H_j|^2}{N_o}\right)$$

where $p_j, |H_j|^2$, and $N_o$, respectively are, the allocated power, the channel gain, and the

AWGN noise variance. The Shannon capacity in equation $r_j = \log_2\left(1 + p_j \dfrac{|H_j|^2}{N_o}\right)$ is an

upper bound that asymptotically approaches the transmission rate over wireless channels. Nevertheless, this upper bound is hard to achieve in practice especially in the network under consideration where adaptive modulation and coding (AMC) is adopted. Particularly, WiMAX supports the adaptive constellation size change of M-OAM and M-PSK following the channel gains changes. From a resource allocation perspective, given the required $P_b$ and channel gains, the allocated power $p_j$ and transmission rate $r_j = \log_2 M$, where $M$ denotes the modulation level, can be adapted. This adaptation is performed by inverting the modulation schemes' $P_b$ approximation functions. The exact approximations of the M-QAM and M-PSK $P_b$, respectively are, given by:

$$P_b \approx \frac{4}{\log_2 M} Q\left(\sqrt{\frac{3p_j \dfrac{|H_j|^2}{N_o} \log_2 M}{M-1}}\right)$$

$$P_b \approx \frac{2}{\log_2 M} Q\left(\sqrt{2p_j \frac{|H_j|^2}{N_o} \log_2 M} \sin\left(\frac{\eth}{M}\right)\right)$$

Equations $P_b \approx \dfrac{4}{\log_2 M} Q\left(\sqrt{\dfrac{3 P_j \dfrac{|H_j|^2}{N_0} \log_2 M}{M-1}}\right)$ and $P_b \approx \dfrac{2}{\log_2 M} Q\left(\sqrt{2 p_j \dfrac{|H_j|^2}{N_0} \log_2 M} \sin\left(\dfrac{\delta}{M}\right)\right)$ are

inverted to obtain the constellation size and power adaptation for a specific $P_b$. However-
er, the $Q(\cdot)$ function cannot be easily inverted in practice, because numerical inversions
are necessary. Alternatively, the exact approximation can be written in a form that is
easy to invert. Because both modulation schemes are special cases of the M-ary modulation techniques, equations above can be written as:

$$p_b \approx c_1 \exp\left[\dfrac{-c_2 p_j \dfrac{|H_j|^2}{N_0}}{2^{c_3 r_j} - c_4}\right]$$

Where $c_1 = 0.2, c_2 = 1.5, c_3 = 1$, and $c_4 = 1$ for M-QAM and $c_1 = 0.05$, $c_2 = 6$, $c_3 = 1.9$, and

$c_4 = 1$ for M-QPSK. By assuming "=" instead of "$\approx$" in equation $p_b \approx c_1 \exp\left[\dfrac{-c_2 p_j \dfrac{|H_j|^2}{N_0}}{2^{c_3 r_j} - c_4}\right]$
and solving for M, we obtain:

$$M = c_3 \sqrt{\left(\dfrac{c_2}{-\ln\left(\dfrac{p_b}{c_1}\right)} p_j \dfrac{|H_j|^2}{N_0} + c_4\right)}$$

The adaptive modulation transmission rate as a function of Pb can be obtained by substituting equation above in $r_j = \log_2 M$:

$$r_j = \dfrac{1}{c_3} \log_2\left(c_4 \dfrac{c_2}{-\ln\left(\dfrac{p_b}{c_1}\right)} p_j \dfrac{|H_j|^2}{N_0}\right)$$

Note that the transmission rates equations $r_j = \dfrac{1}{c_3} \log_2\left(c_4 \dfrac{c_2}{-\ln\left(\dfrac{p_b}{c_1}\right)} p_j \dfrac{|H_j|^2}{N_0}\right)$ and

$r_j = \log_2\left(1 + p_j \dfrac{|H_j|^2}{N_0}\right)$ are similar. Thus, a resource allocation scheme that maximizes one of them maximizes the other. This result broadens the applicability of resource allocation schemes to networks that adopt different modulation schemes.

## MAC Sublayer

Resource allocation is one of the major tasks of MAC because the medium access mechanisms of MAC directly affect the spectrum and power utilization.

Despite the advantages of OFDM in mitigating the channels's impairments as mentioned before, underutilization of transmitter power and network subcarriers is its disadvantage. When an OFDM transmitter accesses the channel in a time division manner, e.g., time division multiple access (TDMA), the transmitter is forced to transmit on all available subcarriers $N_{sc}$, although it may require a less number of subcarriers to satisfy its transmission rate requirement. Consequently, the transmitter power consumption increases as the number of subcarriers increases. This disadvantage motivates the development of a PHY technology where transmitters are multiplexed in time and frequency, i.e., OFDMA. In such a technology, the users are exclusively assigned a subset of the network available subcarriers in each time slot. The number of both time slots and subcarriers can be dynamically assigned to each user; this is referred to as dynamic subcarrier assignment (DSA) which introduces multiuser diversity. The multiuser diversity gain arises from the fact that the utilization of given resources varies from one user to another. A subcarrier may be in deep fading for one user while it is not for another user (e.g., the same subcarrier for user O). Allocating this particular subcarrier to the user with higher channel gain permits higher transmission rate. To achieve multiuser diversity gain, a scheduler at the MAC sublayer is required to schedule users in appropriate frequency and time slots.

Point-to-multipoint (PMP) as well as mesh topologies are supportedin the IEEE 802.16 standard. PMP mode operations are centrally controlled by the BS, but mesh mode can be either centralized or decentralized, i.e., distributed. In a centralized mesh, a mesh BS (a node that is directly connected to the backbone) coordinates communications among the nodes. Decentralized mesh is similar to multihop adhoc networks in the sense that the nodes should coordinate among themselves to avoid collision or reduce the transmission interference.

In PMP mode, the uplink (UL) channel, transmissions from users to BS, is shared by all users, i.e., UL is a multiple access channel. On the other hand, downlink (DL) channel, transmissions from BS to subscriber stations (SSs), is a broadcast channel. The duplexing methods of UL and DL include time division duplexing (TDD), frequency division duplexing (FDD), and half-duplex FDD (HFDD). Unlike PMP mode, there is no clear UL and DL channel defined for mesh mode.

An outstanding feature of WiMAX is heterogeneous traffic support over wireless channels. IEEE 802.16 provides service for four traffic types known as service flows. The mechanism of bandwidth assignment to each SS depends on the QoS requirements of its service flows. The service flows and their corresponding bandwidth request mechanisms are as follows:

- Unsolicited grant service (UGS): This service supports constant bit rate traffic. Bandwidth is granted to this service periodically or in case of traffic presence by the BS.

- Real-time polling service (rtPS): This service has been provided for real-time service flows with variable-size data packets issued periodically. rtPS flows can send their bandwidth request to the BS after being polled.

- Non-real-time polling service (nrtPS): This service is for non-real-timetraffic with variablesize data packets. nrtPS can gain access to the channel using monocast or multicast polling mechanisms. Upon receiving a multicast polling, the nrtPS service can take part in a contention in the bandwidth contention range.

- Best effort (BE) service: This service provides the minimum required QoS for non-real-time traffic. The channel access mechanism of this service is based on contention.

## Centralized Subcarrier and Power Allocation Schemes

In a network with a centralized resource allocation scheme, the BS allocates OFDM subcarriers and power to the users in both UL and DL based on the perfect knowledge of CSI. The users estimate the CSI and report it to the BS at the beginning of each MAC frame interval. It is assumed that the estimation error is negligible and the CSI remains constant during the next frame duration . The BS assigns the resources, subcarrier, and power based on the CSI and broadcast the allocation vector on a signaling channel at the beginning of the next MAC frame transmission. The main difference between UL and DL resource allocation is the power limitation. In DL, the maximum allocated power is limited by the BS power, $P_{BS}$. However, in UL allocated power to the subcarriers of each user is limited by the user's device transmission power which is assumed to be totally devoted to transmission unless it causes an unacceptable interference among neighbors.

## Problem Formulation

DL resource allocation is usually modeled as an optimization problem whose objective function and constraints are determined based on the users' requirements and network specifications. Depending on the definition of the objective functions, different utilization performance are expected. Resource allocation algorithms are available in the literature focus on two general objectives; either data rate maximization or power minimization subject to constraints based on the network model. Using a general objective function of rate, $F(r)$, we present a model for the subcarrier and power allocation optimization problem constrained by the BS maximum power.

Mixed integer nonlinear programming (MINLP) model is appropriate where a discrete network structure and continuous parameters are simultaneously formulated. In DL, the subcarrier and power allocation problem is usually addressed as an MINLP optimization problem. The feasible region of the MINLP model contains integer variables representing subcarriers allocated to the users and continuous variables representing the power allocated to the subcarriers. The network parameters used in the optimization model are given in table.

To show the subcarrier assignment to user $i$, a $K \times 1$-vector $c_i$ of binary variables, called the subcarrier allocation vector of user $i$th, is defined with elements as follows:

$$c_{ij} = \begin{cases} 1 & \text{if jth subcarrier is assigned to ith user} \\ 0 & \text{otherwise.} \end{cases}$$

Each user can be allocated several subcarriers, but each subcarrier is exclusively allocated to one user. A subcarrier may not be assigned to any user due to its severe channel gain. This constraint is mathematically shown by:

$$\sum_{I=1}^{U} C_{ij} \leq 1 \, \forall j \in \mathcal{K}$$

Table: Notions Description.

| Notion | Description |
|---|---|
| U | Total number of users in the network |
| K | Total number of the OFDM subcarriers in the network |
| $\mathcal{K} := \{1, 2, ...,K\}$ | The set of subcarriers |
| $\mathcal{U} := \{1, 2, ...,U\}$ | The set of users |
| i | ith user |
| j | jth subcarrier |
| $\left|H_{ij}\right|^2$ | The channel gain of the ith user on the jth subcarrier |
| $N_o$ | AWGN noise variance |
| $P_{ij}$ | Allocated power to the ith user on the jth subcarrier |
| $r_{ij}$ | Allocated rate to the ith user on the jth subcarrier |
| PBS | BS total power budget |
| $R_{min}^i$ | Minimum service rate requirement of the ith user |

If the $j$th subcarrier is not assigned to the $i$th user, the allocated power to the $i$th user on the $j$th subcarrier must be zero. Therefore, for every user $i = 1, ..., U$ and every subcarrier $j = 1, ..., K$, we must have:

If $c_{ij} = 0$ then $p_{ij} = 0$.

We include this restriction in the model through the following equation:

$$p_{ij} \leq P_{BS} c_{ij} \quad \forall i \in U, \ \forall j \in \mathcal{K}.$$

Note that, if $c_{ij} = 0$, the equation $p_{ij} \leq P_{BS} c_{ij} \ \forall i \in U, \ \forall j \in \mathcal{K}$ implies $p_{ij} \leq 0$ that along with the non-negativity constraint $p_{ij} \geq 0$ yields $p_{ij} = 0$ and satisfies the assumption equation, if $c_{ij} = 0$ then $p_{ij} = 0$. When $c_{ij} = 1$, equation $p_{ij} \leq P_{BS} c_{ij} \ \forall i \in U, \ \forall j \in \mathcal{K}$ is reduced to the redundant constraint $p_{ij} \leq PBS$, because we have the following equation to assure that the total allocated power to all subcarriers in each time slot is limited by PBS:

$$\sum_{i=1}^{U} \sum_{j=1}^{K} c_{ij} p_{ij} \leq P_{BS}$$

In the presence of the set of equation $p_{ij} \leq P_{BS} c_{ij} \ \forall i \in U, \ \forall j \in \mathcal{K}$, that guarantees the restriction equation, if $c_{ij} = 0$ then $p_{ij} = 0$, the variables $c_{ij}$ can be removed from the equation $\sum_{i=1}^{U} \sum_{j=1}^{K} c_{ij} p_{ij} \leq P_{BS}$ as follows:

$$\sum_{i=1}^{U} \sum_{j=1}^{K} p_{ij} \leq P_{BS}$$

The transmission rate to each user depends on the number and index of allocated subcarriers to the user and allocated power to each subcarrier. If continuous rate adaptation is assumed, the approximate rate of the $i$th user, $r_i$, can be obtained by either

$$\text{equation } r_j = \log_2 \left( 1 + p_j \frac{|H_j|^2}{N_o} \right) \text{ or equation } r_j = \frac{1}{c_3} \log_2 \left( c_4 \frac{c_2}{-\ln\left(\frac{p_b}{c_1}\right)} p_j \frac{|H_j|^2}{N_o} \right) \text{ as follows:}$$

The heterogeneous traffic in the network inquires different QoS. The minimum service rate requirement of the $i$th user, $R^i_{min}$, is guaranteed through the following equation:

$$r_i \geq R^i_{min} \quad \forall i \in \mathcal{U}$$

Respecting the equation listed above and the chosen objective function, $F_i(r_i)$, for user $i$, the resource allocation optimization problem can be modeled as follows:

$$(P_1): \max_{cij, \, pij} \sum_{i=1}^{U} F_i(r_i)$$

$$\text{s.t. } r_i = \sum_{j=1}^{K} r_{ij} \quad \forall i \in \mathcal{U},$$

$$r_i \geq R^i_{min} \quad \forall i \in \mathcal{U},$$

$$\sum_{i=1}^{U}\sum_{j=1}^{K}p_{ij}\le P_{\text{BS}}$$

$$\sum_{i=1}^{U}c_{ij}\le 1$$

$$\sum_{i=1}^{U}c_{ij}\le 1 \quad \forall i \in \mathcal{K}$$

$$0 \le p_{ij} \le P_{\text{BS}}c_{ij} \quad \forall i \in \mathcal{U}, \forall j \in \mathcal{K}$$

$$c_{ij}\in\{0,1\} \quad \forall i \in \mathcal{U}, \forall j \in \mathcal{K}$$

Problems $(P_1)$ is an MINLP problem. Solving MINLP problems can be very challenging, due to the combination of mixed integer and nonlinear program difficulties in MINLP problems. It is proved in Ref. that the integer variables in problem $(P_1)$ are redundant and can be eliminated. Accordingly, a nonlinear programming (NLP) model is proposed that unifies the subcarrier and power allocation in a rate allocation problem. Equation $\sum_{i=1}^{U}c_{ij}\le 1 \quad \forall i \in \mathcal{K}$, defined below, is replaced by Equation for this purpose. For every $\hat{i} \in U$ and for every $j \in \mathcal{K}$:

$$r_{\hat{i}j}\cdot r_{ij} = 0 \quad \forall j\in\mathcal{K},\forall i\in\mathcal{U},\forall \hat{i}\in\mathcal{U}, i = \hat{i}$$

It is proved that equation $r_{\hat{i}j}\cdot r_{ij} = 0 \quad \forall j\in\mathcal{K},\forall i\in\mathcal{U},\forall \hat{i}\in\mathcal{U}, i =\hat{i}$, the same as equation $\sum_{i=1}^{U}c_{ij}\le 1 \quad \forall i \in \mathcal{K}$, guarantees exclusive rate allocation to the $i$th user on the $j$th subcarrier and the optimal value of problem $(P_1)$ equals the optimal value of problem $(P_2)$ stated as follows:

$$(P_2): \max_{r_{ij}} \sum_{i=1}^{U}F_i(r_i)$$

$$\text{s.t. } r_i = \sum_{j=1}^{K}r_{ij} \quad \forall i \in \mathcal{U},$$

$$r_i \ge R_{\min}^{i} \quad \forall i \in \mathcal{U},$$

$$\sum_{i=1}^{U}\sum_{j=1}^{K}\frac{1}{r_{ij}}\left(2^{r_{ij}}-1\right)\le P_{\text{BS}}$$

$$r_{\hat{i}j}r_{ij} = 0 \quad \forall i\in\mathcal{U}/\{\hat{i}\} \,\forall j\in\mathcal{K}$$

$$0 \le r_{ij} \quad \forall i \in \mathcal{U}/\{\hat{i}\} \; \forall j \in \mathcal{K}$$

The new model will be easier to deal with, both by optimal/suboptimal and heuristic approaches. As the objective function is continuous over the range of allocated rates and the feasible region is closed and bounded, the extreme value theorem (Weierstrass Theorem) implies that problem $(P_2)$ has global optimal solution(s). Extreme value theorem: Let $f$ be a continuous real-valued function whose domain, $D_f$, is bounded and closed. Then, there exist $x_1$ and $x_2$ in $D_f$ such that:

$$f(x_1) \le f(x) \le f(x_2) \; \forall x \in D_f.$$

## Decentralized Subcarrier and Power Allocation Schemes

The distributed infrastructure of WiMAX mesh and relay networks, and the need for reducing communication overhead between the BS and network nodes are the motivations behind proposing decentralized resource allocation schemes. Potentially, either no central controller exists or it does not influence the allocation decision in a decentralized resource allocation. These facts make the decentralized schemes more scalable.

Resource allocation in decentralized networks is essentially different from centralized one. In centralized schemes, the BS collects CSI from all users, allocates the subcarriers or power to the users, and informs the users of allocated resources. On the other hand, users may not need to know the CSI of the other parties in decentralized schemes. Besides, the parties may not be aware of the decision of each other, so a collision is probable. Accordingly, in each proposed distributed resource allocation scheme for OFDM-based networks, the following questions should be answered:

- How does a user achieve the required CSI? The hidden terminal and exposed node problems are common problems in self-organized and decentralized networks, because it is assumed that no central controller exists to assist in signaling. Besides, the signaling overhead should be reasonable for a practical implementation.

- How should the node coordinate or compete with the other nodes to attain the resources? A node does not know the requirements of the other nodes, at each instant, so competence or coordination is a "must" for a node which wants to start capturing the resources.

An interference aware subchannel allocation scheme that overcomes the drawbacks of decentralized schemes, i.e., hidden and exposed node problems is proposed. As the scheme uses updated CSI at the beginning of each MAC frame, the channel does not need to be assumed time-invariant over multiple MAC frames. The scheme is appropriate for OFDM-TDD networks. The MAC frame is divided into mini-slots; at the first mini-slot of each MAC frame the ongoing receiver nodes broadcast a busy signal to inform the respected transmitters of the quality of the allocated subchannel. The inactive nodes does not send any busy signal. The ongoing transmitter nodes listen to the busy signal and adapt their subcarrier allocation to their specific receiver nodes according to the information on the busy signal. Also, a node that wants to start transmission listens to the busy signal and chooses the subcarriers that are not interfered by ongoing transmissions and their interference. In other words, it selects those subcarriers with a received busy signal power less than a threshold. The advantages of the scheme are as follows:

- Signaling overhead is low compared to other radio resource allocation schemes.

- The co channel interference is reduced significantly since the scheme is interference aware.

- Full frequency reuse is possible.

A decentralized power allocation problem for a cooperative transmission is formulated. The objective is to minimize the total transmission power of all users subject to providing minimum rate requirement of each user. The network model uses a time division multiple access with the OFDM multiplexing (TDMA-OFDM), so only one user accesses the total OFDM subcarriers in each time slot. The user may allocate some of the subcarriers to transmit its traffic and the rest of subcarriers to relay the other users' traffic to minimize total transmission power. In other words, the resource allocation problem determines if a user should cooperate with other users, and in case of cooperation it determines the subcarriers and power allocation. A cooperative user may be allocated more power than a non cooperative user, because its location and channel gain allows it to cooperate with other users. However, the total network power is minimized by cooperation among users. A decoupled subcarrier and power allocation scheme is proposed to solve the problem.

Assuming an access point in the network, proposes a distributed decision-making scheme for the resource allocation. Each user measures its CSI upon receiving a beacon signal from the access point. The subcarriers are divided into several groups (equal to the number of users), and the approximate channel gain of each group for each user is estimated. Then, the users contend with each other to achieve the group with the best channel quality of their own. A backoff mechanism is proposed to avoid collision. Each user start contending for the best group of subcarriers after a backoff time which is proportional to the best group gain for that user. After contention, the access point informs the users of the winners which can transmit in the next transmission interval. A frame is divided into three subframes, contention, acknowledge, and transmission sub frames. This scheme is actually a distributed resource allocation scheme for a centralized network that aims to reduce the signaling overhead and processing task of the access point.

A collaborative subcarrier allocation using the swarm intelligent is proposed in. The scheme relies on the central controller to achieve the updated information of available subcarriers and the highest and lowest demands for each subcarrier. In other words, the nodes negotiate with the central controller iteratively in a negotiation phase. In each iteration, the nodes inform the central controller of their demand for a specific subcarrier. Then, the central controller broadcast a message indicating the highest and lowest demands for each subcarrier. Based on the feedback messages that occur several times in each negotiation phase, the nodes intelligently decide upon subcarriers.

A comparison between the capacity performance of OFDM-TDMA and OFDM-FDMA in a two-hop distributed network is performed. The time frame is divided by two in OFDMTDMA. In each half of the time frame, all subcarriers are devoted for transmission over one hop. A power allocation algorithm is proposed to maximize the end-to-end capacity subject to the overall transmit power constraint of the two hops. For OFDM-FDMA, the subcarriers are assigned to the two hops without overlapping and a joint subcarrier and power allocation algorithm is proposed. The simulation performance results of the algorithms show that OFDM-FDMA achieves a higher end-to-end capacity than OFDM-TDMA.

## References

- OFDM-Modulation-Technique-its-Applications-A-Review-327043286: researchgate.net, Retrieved 16 April, 2019

- Basics-techniques, radio-multicarrier-modulation: electronics-notes.com, Retrieved 08 July, 2019

- Roque, D.; Siclet, C. (2013). "Performances of Weighted Cyclic Prefix OFDM with Low-Complexity Equalization" (PDF). IEEE Communications Letters. 17 (3): 439–442. doi:10.1109/LCOMM.2013.011513.121997

- IEEE standard for local and metropolitan area networks part 16: Air interface for fixed broadband wireless access systems, October 2004

- Multicarrier Techniques for 4G Mobile Communications, Artech House, 2003

# Applications

WiMAX is used for a wide-range of applications such as broadband connections, cellular backhaul, hotspots, e-learning, wider metropolitan area network access to users, etc. This chapter closely examines the different applications of WiMAX for an extensive understanding of the subject.

## MEDICAL CONSULT-BASED SYSTEM FOR DIAGNOSIS ON WIMAX TECHNOLOGY

WiMAX (Worldwide Interoperability for Microwave Access) is a high-speed wireless broadband technology that was developed on the IEEE 802.16 standards and has developed to support mobile usage by setting a new standard IEEE 802.16e, the ability to send signals to spread from point-to-multipoint simultaneous with the ability to work in support NonLine-of-Sight.

At present, heart disease is the most serious disease and its incidence increases with age that patients must be treated constantly monitored closely. Most of the cardiac deaths occur outside of the hospital.

Communicative wearable health systems are recognized as one of the most promising platforms for minimally obtrusive and individualized health services at the point of need. The new generation of biomedical sensors presents a large spectrum bandwidth allowing new measurements on humans and new approaches for diagnosis, ambulatory healthcare, and care at the point of need, any time.

The electrocardiogram (ECG) is very important for diagnosis Care to prevent patients receiving life threatening. In progress has been made in the development of a remote monitoring system for ECG Signals, the deployment of packet data services over telecommunication network with new applications The tele-transmitting and receiving of ECG Signals is the main problem for medical-diagnosis and monitoring. This system is provided remote monitoring of one or several patients wearing portable devices equipped with a PDA-based portable wireless ECG monitoring for physician personal area networks wireless connectivity based on different technologies.

Recently, the fast development of mobile technologies, including increased communication bandwidth and miniaturization of mobile terminals, has accelerated developments in the field of mobile telemedicine. Wireless patient monitoring systems not only increase the mobility of patients and medical personnel but also improve the quality of health care. The IEEE 802.16e mobile WiMAX system describes a wireless broadband technology that can support mobile users. Mobile WiMAX is the wireless network which can connect Internet in high data rate anywhere and anytime. Video surveillance systems are very effective and important for security management. Combining traditional cameras and network video technology together, the IP camera can compress and deliver live video clips to the Internet without using a PC so that people can remotely browse and watch the protected area activities. Hence, the IP cameras are key devices of the surveillance systems.

## Structure of the System

Figure shows the structure of overall system. The system consists of four subsystems namely:

- Portable ECG Transceiver check status of patient and send the biosignal with ZigBee/IEEE RF module.

- IP video camera pans video of patient.

- WiMAX transceiver sends ECG signals and panning video camera of the patient to the physicians via WiMAX technology.

- Personal monitor consist application on physician's PDAs or notebook for monitor information system, analysis data and view panning video camera of patient.

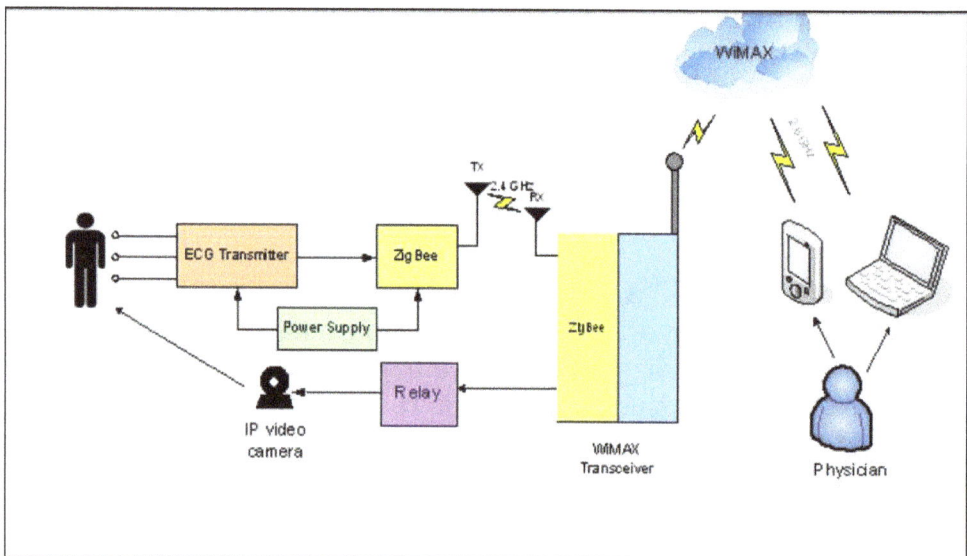

Structure of overall system.

## System Implementation

Physicians can use PDA connect to the IP video camera by the web browser and monitoring ECG signals by the application on physician's PDA via WiMAX technology.

## Portable ECG Transceiver

The ECG signals acquired from electrodes must be amplified by amplifier and eliminate range of frequency of unwanted signal left before to the microcontroller. Then, the ECG signals from microcontroller will be connects via ZigBee/IEEE technology which is XBee-PRO OEM RF on IEEE 802.15.4 standards and support the unique needs of low-cost, low-power wireless sensor networks. Figure illustrates essential blocks for portable ECG transmitter and figure shows transmit ECG signals by ZigBee/IEEE technology.

## IP Video Camera

At ZigBee's receiver, the abnormal ECG signals of patient will lead to trig the relay for open the IP video, camera. If the relay is trigged, the IP video camera that connects to the relay will be opened.

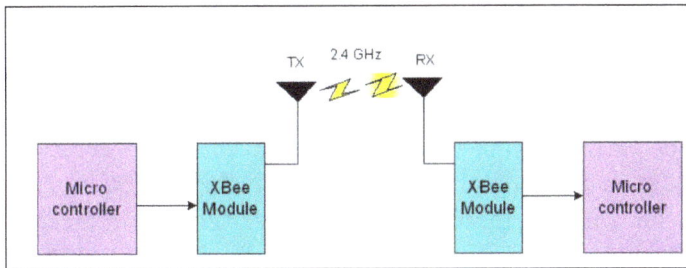

ZigBee/ IEEE technology.

Circuit board of ECG amplifiers.

## WiMAX Transceiver

WiMAX refers to interoperable implementations of the IEEE 802.16 wireless-networks standard, in similarity with Wi-Fi, which refers to interoperable implementations of the IEEE 802.11 wireless LAN standard. The bandwidth and range of WiMAX provide portable mobile broadband connectivity across cities and ecountries through a variety of devices and a wireless alternative to cable and DSL for last mile broadband access.

There are numerous devices that provide connectivity to a WiMAX network. These are known as the subscriber unit (SU) such as Residential Subscriber Unit, Outdoor Subscriber Unit, Mobile Subscriber Unit and PC Card CPE (customer-premises equipment). There is an increasing focus on portable units, this includes handsets (similar to cellular smart phones), PC peripherals (PC Cards or USB dongles), and embedded devices in laptops, which are now available for Wi-Fi services. It is notable that WiMAX is more similar to Wi-Fi than to 3G cellular technologies. WiMAX and Wi-Fi are related to wireless connectivity and Internet access.

WiMAX Base Transceiver Station (WiMAX BTS) delivers flexible, high-speed connectivity for fixed and nomadic wireless broadband access. It will keep costs down as your system needs change by supporting a highly scalable system architecture for both rural and metropolitan areas. In conjunction with the Residential Subscriber Unit, Outdoor Subscriber Unit, Mobile Subscriber Unit, and PC Card CPE, the BTS incorporates automatic adaptive modulation to maximize the capacity of the airlink over a wide variety of configurations and propagation conditions.

WiMAX operates on the same general principles as WiFi, it sends data from one computer to another via radio signals. A computer (either a desktop or a laptop) equipped with WiMAX would receive data from the WiMAX transmitting station, probably using encrypted data keys to prevent unauthorized users from stealing access. The fastest WiFi connection can transmit up to 54 megabits per second under optimal conditions. WiMAX should be able to handle up to 70 megabits per second. The biggest difference isn't speed; it's distance. WiMAX outdistances WiFi by miles. WiFi's range is about 100 feet (30 m). WiMAX will blanket a radius of 30 miles (50 km) with wireless access. The increased range is due to the frequencies used and the power of the transmitter.

WiMAX can be used for wireless networking in much the same way as the more common WiFi protocol. WiMAX is a second-generation protocol that allows for more efficient bandwidth use, interference avoidance, and is intended to allow higher data rates over longer distances. In practical terms, WiMAX would operate similar to WiFi but at higher speeds, over greater distances and for a greater number of users. WiMAX BTS use channel bandwidth at 2.6 GHZ in the range of 3-6 MHz, operating in mode time division duplex (TDD), RF output power equal to 2 watts, RF input impedance and RF output impedance are 50 ohms, Power Requirements about 36 to 60 vdc and Average Power Consumption is 100 watts.

## Personal Monitor

A physician access by remote monitoring client, and can request information of patient by PDA or notebook. These must be install ECG monitor software for get patient's information and connected to hospital server including view panning video camera of patient through the web browser via WiMAX technology.

# WIMAX TECHNOLOGY FOR E-LEARNING SOLUTIONS

E-learning is widely used today on different educational levels: continuous education, company trainings, academic courses, etc. The educational process is a complex service which involves a producer and a consumer.

The e-learning solutions can be implemented in inaccessible locations (isolated localities and communities) using several mobile technologies: GSM/UMTS, WiMAX etc.

One of the promising technologies is WiMAX. It provides high speed data transfer and can be used in such locations, where other communications technologies are not available.

Usually, e-learning systems are developed as distributed applications, but this is not necessary so. The architecture of a distributed e-learning system includes software components, like the client application, an application server and a database server the necessary hardware components (client computer, communication infrastructure and servers).

E-learning clients have to be developed having in mind users' requirements, several studies in this area being made for mobile clients.

Figure presents an architecture of an e-learning solutions based on WinMAX technology.

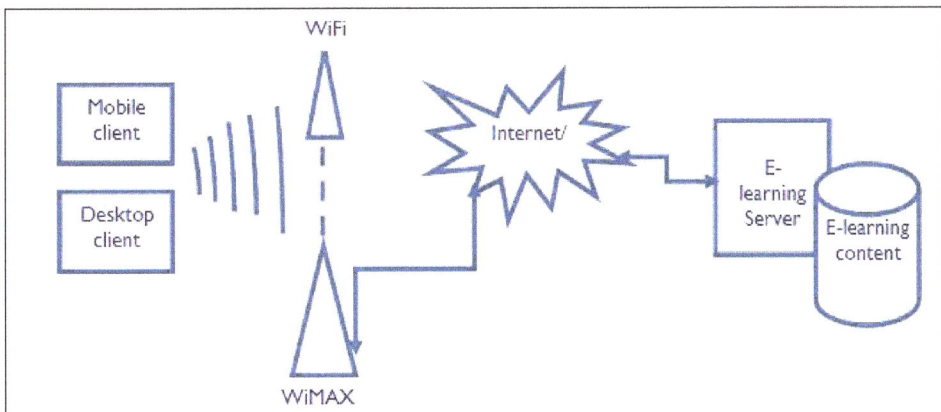

An e-learning solution using WiMAX.

The e-learning clients connect to the e-learning platform by using WiFi networks or connecting directly to the WiMAX network (if appropriate hardware is available).

The e-learning solution can be integrated into cloud architecture. A very big concern is related to the data security because both the software and the data are located on remote servers that can crash or disappear without any additional warnings. Even if it seems not very reasonable, the cloud computing provides some major security benefits for individuals and companies that are using/developing e-learning solutions, like the following:

- Improved improbability: It is almost impossible for any interested person (thief) to determine where is located the machine that stores some wanted data (tests, exam questions, results) or to find out which is the physical component he needs to steal in order to get a digital asset.

- Virtualization: Makes possible the rapid replacement of a compromised cloud located server without major costs or damages. It is very easy to create a clone of a virtual machine so the cloud downtime is expected to be reduced substantially.

- Centralized data storage: Losing a cloud client is no longer a major incident while the main part of the applications and data is stored into the cloud so a new client can be connected very fast. Imagine what is happening today if a laptop that stores the examination questions is stolen.

- Monitoring of data access becomes easier in view of the fact that only one place should be supervised, not thousands of computers belonging to a university, for example. Also, the security changes can be easily tested and implemented since the cloud represents a unique entry point for all the clients.

Wireless data communication can be easily monitored, so high security need to be assured by using specific standards. For example, if WiFi is used, it is recommended to use WPA2 standard combined with other WiFi security techniques.

WiMAX is an appropriate solution for e-learning platforms when the trainee location is isolated. One of the main concerns is related to security.

Currently there are no efficient solutions to prevent the attacks at the physical layer of a WiMAX network but, despite of all issues and threats, WiMAX is considered to be a secure network that provides:

- Strong user authentication.

- Access control.

- Data privacy.

- Data integrity.

- Using sophisticated authentication and encryption technology.

## MOBILE BACKHAUL

Mobile backhaul refers to the transport network that connects the core network and the RAN (Radio Access Network) of the mobile network. Recently, the introduction of small cells has given rise to the concept of fronthaul, which is a transport network that connects the macrocell to the small cells. Whilst mobile backhaul and fronthaul are different concept, the term mobile backhaul is generally used to encompass both concepts. Furthermore, innovations to reduce demand on mobile backhaul sometimes involve architectural changes in the antenna (also referred to as radio unit in 4G) and the controller (also referred to as digital unit in 4G). Therefore, the components labelled in red in the figure below will be covered for backhaul demand case studies within the GSMA Future Networks Network Economics.

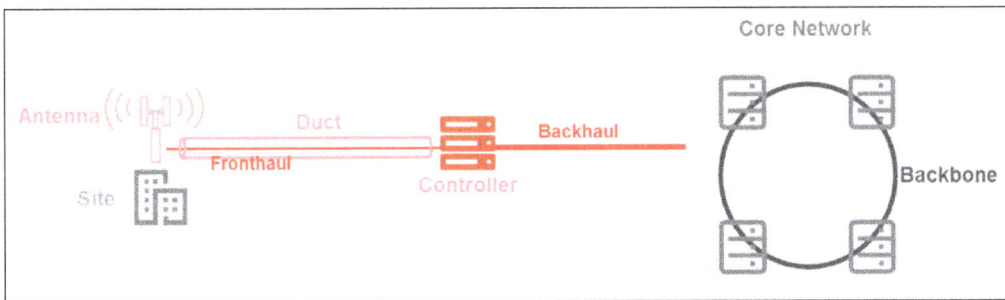

Mobile network and the scope of mobile backhaul.

### Success of LTE and Growing Importance of Mobile Backhaul

Wireless and fixed-line backhaul infrastructure is an essential component of the mobile telecommunications network. Mobile networks are ubiquitous and support a mix of voice, video, text and data traffic originating from and terminating to mobile devices. All of this traffic must be conveyed between the mobile cellular base stations and the core network. The success of 4G Long-Term Evolution (LTE) has placed even greater challenges on mobile operators as they strive for more network capacity, latency reduction, and the need to deliver an enhanced user experience. Notwithstanding the success of LTE (as of May 2019, there were 729 LTE operators in more than 221 countries), there are also a number of new challenges on the horizon that will impact on the MNO's backhaul network infrastructure.

In the era of 5G, where a network will be densified and more stringent requirement will be imposed, mobile backhaul will become even more important. Given the limitations governed by laws of physics, the mobile backhaul will add much more pressure on mobile operators' cost. Whilst mobile operators will be able to unlock new business opportunities with 5G, the costs need to be optimised for the operators to sustainably reap the benefits of 5G.

## Challenges in Mobile Backhaul

There are a number of market trends that result in new challenges and requirements that must be met by the backhaul infrastructure of MNO networks.

## Evolution of LTE

There are a number of technical innovations occurring on LTE, which is known as LTE-Advanced Pro or 4.5G which enable enhancements such as improved peak bandwidth and greater energy efficiency for IoT connections. The peak bandwidth of 4.5G is around 1Gbps which is 8-10x higher than standard LTE, and will enable (inter alia) support of video traffic at 4K resolution to mobile devices.

## Emergence of 5G

By the end of 2018, there have already been some 5G deployments for FWA (Fixed Wireless Access) but the first mainstream 5G mobile services are expected to commence in 2020. In early 2018, there were 113 operators in 56 countries doing 5G trials and it is anticipated that by 2024, 50 countries will have in-service 5G networks. The 5G network will comprise both NR (New Radio) as well as a new 5G Core Network (5GC). The advent of NR offers a leap in bandwidth speeds in comparison to 4G via the utilisation of higher frequency spectrum. There will be 3 separate 5G bands initially, namely Sub-1 GHz, 1-6 GHz and above 6GHz. It is expected spectrum bands above 24GHz will be agreed upon at the WRC in 2019, which includes 26GHz and 40GHz bands. The higher frequencies enable wider channel bandwidths at the access but also result in smaller cell sizes. Both have implications for backhaul.

## 5G Network Slicing

5G Network Slicing. One of the key features of the 5G Network is the concept of "network slicing" whereby the physical network infrastructure can be partitioned into bespoke logical networks ("slices") in the RAN and 5G core which are targeted to the needs of a specific application or use case. Slicing will also impact on the backhaul network and will also facilitate sharing of infrastructure to optimise cost (for more information on infrastructure sharing, seehere).

## Subscriber Growth

At the end of 2017, subscriber numbers stood at over 8.1 billion with an annual growth rate of 5.4% year on year. It is estimated that by 2025, the number of subscriptions will be 9.8 billion. In terms of different RAN technologies, LTE subscriptions were at 2.86 billion in 2017 (35.2% of the overall total) and will be at 4.24 billion (43.3% of the overall total) by 2025 whilst 5G subscriptions will be around 850 million by 2025. Furthermore, IoT devices (LTE-M and NB-IOT) will grow from 376 million to 4.2 billion in the 2017-25 timeframe. Therefore, backhaul strategy/evolution must cope with both an increase in subscriptions as well as a large number of those subscriptions being "high bandwidth" users.

## Mobile Data Traffic Growth

The increasing subscriber total plus increased access bandwidth usage of those subscribers results in mobile data traffic increasing at a rate of 28.9% CAGR to reach over 1300 Exabytes [1300 x $10^{18}$ bytes] by 2025. By 2025, 4G and 5G subscribers will represent 55% of subscriptions but will generate 91% of the traffic. There will also be a marked shift in the type of traffic being carried with video streaming increasing from 81 Exabytes (which is just under 50% of total traffic) to 910 Exabytes (which is 70% of total mobile traffic). Per-user traffic per month is expected to grow from 1.7 GBytes to 11.3 GBytes between 2017 and 2025. In some markets,

## Stringent Latency Requirements

Both 5G mission-critical applications and increased video streaming will result in more stringent end-end latency requirements and impact on the backhaul latency budget. For example, an end-end latency cap of 10ms implies a latency across the backhaul that is <1ms – which means that only fibre optic and microwave links will be able to support such low latency requirements. If higher latency backhaul links are deployed (e.g. satellite links), then such backhaul would only carry 2G/3G and non-latency sensitive LTE services.

## Network Densification

The increased demand for mobile broadband results in the number of macrocell sites being estimated to grow globally from 11.1 million to 14.1 million. The new macrocells include both 4G and 5G technologies. This results in extra traffic to backhaul as well as additional challenges due to the smaller cell size for 5G NR.

Furthermore, the growth of LTE traffic resulted in MNOs being increasingly reliant on small cell site deployments. Small cells will be even more essential for 5G. Small cells can be deployed outdoors or indoors and include low power microcells, femtocells and picocells. They can be deployed on private or public infrastructure in the urban environment (e.g. rooftops, street poles etc.). In the period 2017-25, the number of small cells is expected to increase globally from 0.71 million to 4.3 million. From a backhaul point of view, there is a need to be able to carry traffic from many more cell sites in a scalable, efficient and economic manner.

## Technology Choices for Mobile Backhaul

There are a number of technical solutions used by Mobile operators for backhaul, including both wireline and wireless solutions.

## Copper-line

Copper-based backhaul was the primary backhaul technology for 2G/3G. At the heart of copper-based backhaul is the T1/E1 protocol, which supported 1.5 Mbps to 2 Mbps. This bandwidth can be boosted by using DSL over the copper pair as well as so-called copper-bonding (i.e. multiple copper pairs are bonded together). For example, having up to 12 bonded pairs can provide over 150 Mbps downlink capacity over 1.5 km. Nevertheless, copper lines do not scale easily to provide adequate bandwidth at a distance above a few hundred meters to support LTE broadband usage and 5G traffic scenarios will prove particularly challenging for mobile service providers. In bonded configurations, as monthly costs increase linearly with bandwidth requirements. Therefore, as bandwidth requirements become more onerous, copper-based backhaul has become an infrequently used solution and operators are increasingly preferring fibre-optic where available (e.g.. in city centres). For all that, DSL is still an option for mobile backhaul for indoor small cells, in-building HetNets, and public venue small cell networks.

## Fibre-optic

This technology is the mainstay wired backhaul in MNO networks and second overall only to microwave backhaul. Even though fibre has significant inherent bandwidth carrying capability, several additional techniques can be used to offset any bandwidth constraints and essentially rendering the fibre assets future-proof. These techniques include Wavelength Division Multiplexing (WDM) technology which enables multiple optical signals to be conveyed in parallel by carrying each signal on a different wavelength or colour of light. WDM can be divided into Coarse WDM (CDWM) or Dense WDM (DWDM). CWDM provides 8 channels using 8 wavelengths, while DWDM uses close channel spacing to deliver even more throughput per fibre. Modern systems can handle up to 160 signals, each with a bandwidth of 10 Gbps for a total theoretical capacity of 1.6 Tbps per fibre. The traffic generated by LTE has accelerated the demand for Fiber to the Tower (FTTT) and has required Mobile Network Operators (MNOs) to upgrade many aspects of their backhaul networks to fibre-based Carrier Ethernet. The main limitations of fibre are the cost and logistics of deploying fibre (ducts etc.), although the cost of fibre has been decreasing over the last few years (e.g. it now costs circa $70K/km whereas 5 years ago it cost circa $100K/km). Nonetheless, it can still take several months to provision a cell site with fibre optic backhaul. Fibre backhaul was used for 26% of global macrocell backhaul links in 2017, growing to just under 40% by 2025. Fibre will also be the main choice.

## Wireless Backhaul (Microwave)

Despite fibre being the preferred choice for MNOs for 4G/5G backhaul, microwave backhaul is the most used technology due to a combination of its capability and relative ease of deployment (i.e. no need for trenches/ducting) making it a low-cost option that can be deployed in a matter of days. Most MNOs rely heavily rely on microwave backhaul solutions in the 7 GHz to 40 GHz bands, in addition to higher microwave bands such as V-band (60 GHz) and the E-band (70/80 GHz). Backhaul links using the V-band or the E-band are well suited to supporting 5G due to their 10 Gbps to 25 Gbps data throughput capabilities. Microwave can be used in LOS or NLOS mode which makes it ideal to be used in a chain, mesh or ring topologies to enable resilience and/or reach. The main drawback is that microwave backhaul requires a licence, apart from the V-band that is unlicensed and to a lesser extent the E-band which is lightly licenced. It is also possible to combine a low-frequency microwave band with a high-frequency microwave band to achieve high capacity over long distances with enhanced availability.

## LOS vs. NLOS

Historically, most wireless backhaul links have been LOS (Line of Sight) due to the high frequencies being used, as well as the narrow beam widths used. However, over the past 10 years, NLOS (Non-Line of Sight) has become a viable solution that should prove particularly advantageous with a large number of clusters of small cells that MNOs are expected to deploy over the next few years.

LOS backhaul has the advantage of using a highly directed beam with little fading or multi-path dispersion and enables efficient use of spectrum as multiple transceivers can be located within a few feet of each other and use the same frequency to transmit different data streams. On the other hand, it may be difficult to always have an unobstructed path in certain scenarios (e.g. trees, buildings in the way) and the transceiver pair need precise alignment. In the latter case, this can be impacted by "pole tilt" where the alignment is spoiled by movement caused by the wind and a particular concern for small cells. The pole tilt issue gets worse as frequencies increase due to the beam narrowing. For large numbers of small cells (e.g. in a metropolitan hot spot), the cost of backhaul can increase quickly if a number of links are daisy chained together.

NLOS backhaul is much more "plug and play" and so take less time with less skilled labour to set up. NLOS backhaul OFDM technology (Orthogonal Frequency Division Multiplexing) to relay information back to a central base station. NLOS backhaul needs only to be within a range of the receiver unit with OFDM providing a level of tolerance to multi-path fading not possible with LOS. There is a limit to how many small cells can be blanketed by a single NLOS backhaul to ensure each cell has a given QoS as all of the bandwidth is shared between the multiple base stations covered by the central unit. This bandwidth sharing is a disadvantage in that there is an upper limit on the bandwidth available to each base station and calculations need to be re-done if further base stations are added. Frequency planning also needed to avoid interference as the NLOS frequency ranges can also be used for access.

## Satellite Backhaul

Satellite Backhaul is a niche solution for MNOs and used in fringe areas (e.g. remote rural areas) and sometimes as an emergency/temporary measure (e.g. a disaster area or in place of a microwave link whilst waiting for licence approval). This backhaul is used in developing markets and as a complementary role in developed markets. The technology can deliver 150Mbps/10Mbps (downlink/.uplink). However, latency is a challenge as there a round trip delay of circa 500-600ms for a geostationary satellite. LEO (Low Earth Orbit) satellites have tried to address the latency issue (i.e. using a much lower orbit of 1500km versus 36000km and resulting in a one way trip of circa 50ms). However, LEO satellites are not geostationary and thus there is sometimes a need to route traffic via multiple satellites. LEO satellites are also relatively immature technology. Fees are also usage based on satellite links and means that such links need to be monitored and controlled.

## WiFi Backhaul

There is marginal (<1%) use of this technology for macrocell backhaul in some emerging markets. The unlicensed nature of the technology combined with the growing interference from increasing public and private WLANs plus poor transmission ranges severely limits its deployment.

## Market Share and Trends

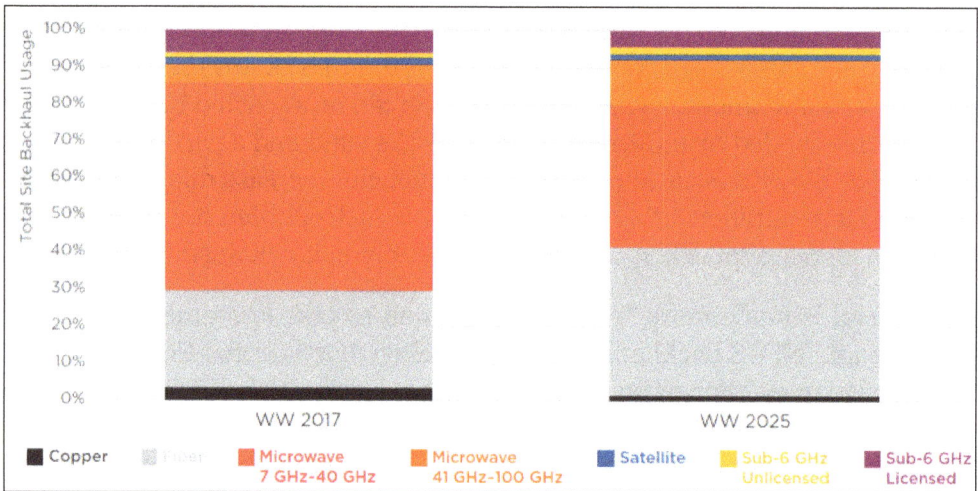

Backhaul technology trend and forecast.

In terms of market share and trends, wireless backhaul (microwave) in the traditional (7-40GHz) range was responsible for nearly 57% of macrocell backhaul links in 2017, diminishing to 45% of macro-cell links by 2025. Microwave links in the 41-100GHz will double from 3.2% to 6.1% in the same period. The shorter range of the latter (<3km) is offset by their increased data throughput and thus make it a suitable technology in urban areas. For small cells, traditional microwave was used for 35.2% of links in 2017 diminishing to 21% in 2025, whilst microwave links in the 41-100 GHz range will grow from 10.4% to 13.1% in the same period.

Fibre based backhaul was responsible for 25.6% of macrocell links in 2017, rising to 39.6% by 2025. Fibre is the market leader for small cell backhaul with 43.2% of the market on 2017, rising to 56.1% in 2025. DSL based backhaul was used for 3.6% of macrocell backhaul in 2017 and this share is expected to decline over the next few years. Satellite comprised 1.9% of backhaul links in 2017 and this will diminish to 1.4% in 2025 (although the overall number will increase), reflecting its niche/complimentary role. These trends are summarized in the figure.

## Alternative Architectures for Mobile Backhaul Optimisation

### MEC (Multi-access Edge Computing)

MEC (Multi-access edge computing) is where computing and intelligence capabilities that were mostly centralized in the core network are provided at the edge of the access network. MEC enables high bandwidth and ultra-low latency access to cloud computing/IT services at the edge to be accessed by applications developers and content providers.

MEC, while incurring a cost to implement core functions at the edge, can provide opportunities to optimise backhaul demand via caching and/or local breakout. Caching reduces the load on mobile backhaul and enhances the customer experience by storing frequently accessed contents in the edge network. Customers can access the contents at a lower latency (with less distance for signal to travel) and backhaul demand is reduced as there is no need to reach further to the external network to obtain the contents. Local breakout also enables the mobile backhaul to be optimised as the contents do not need to travel to the core network and then to the internet. The caveat with local breakout is that the transport network to connect the edge to the internet needs to be in place and therefore won't optimise cost in certain scenarios.

### Cloud RAN

Cloud RAN is where some layers of radio access network are centralized to an edge site rather than at the cell site, which allows some (or all) of the processing capabilities to be focused at the edge site reducing the complexities at the cell site. This architecture is suitable in the small cell era, where only a little space and cost constraint is affordable at the cell site. While the architecture may not be suitable for traditional macrocell base stations as they would need to process significant load of signal transmitted from/received by various radio elements, heterogeneous networks with many small cells would benefit from this architecture.

As shown in the figure, Cloud RAN in its two forms (low-level and high-level splits) significantly reduces complexities and capabilities at the cell site to be concentrated in the edge site. The low-level split is where only the physical layer is processed at the edge site while all the electronics are concentrated in the edge site. This architecture allows easy installation and very low complexity at the cell site but comes at a higher fronthaul cost as baseband signals would need to be transferred. On the other hand, high-level split brings relatively less fronthaul cost but comes with more complexity at the cell site than low-level split.

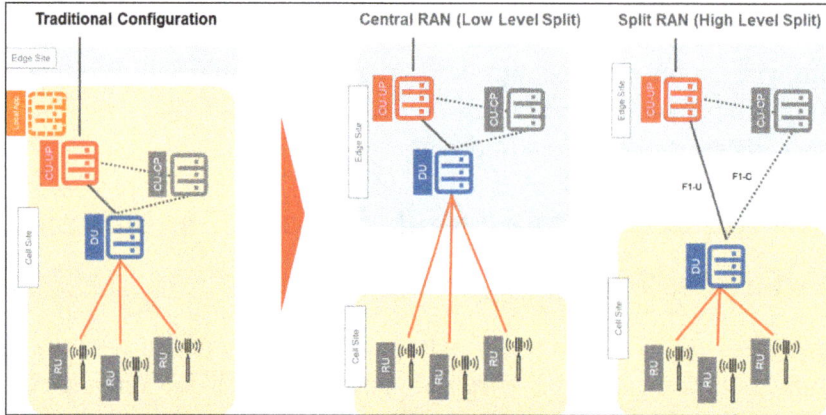

Cloud RAN Architecture.

## Optimal WiMax Backhauling Solutions for WiFi Traffic

The IEEE 802.16 standard or WiMax (Worldwide Interoperability for Microwave Access) is last mile solution for broadband wireless access. With the QoS support explicitly established in WiMax design, service providers are invited to deliver the following services: Unsolicited Grant Service (UGS) for constant bit-rate (CBR) traffic, Real-Time Polling Service (rtPS) for delay-sensitive real-time traffic, Non-Real-Time Polling Service (nrtPS) for delay-tolerant traffic, and Best-Effort Service for less stringent data services applications. This QoS framework allows WiMax Service Providers to offer a variety of applications such as: FTP, HTTP, email, voice, and video by mapping each one, accordingly its QoS demands, in a particular service class.

The WiMAX physical layer (PHY) is based on orthogonal frequency division multiplexing (OFDM), which enables highspeed data, video, and multimedia communications and is used by a variety of commercial broadband systems, including DSL, Wi-Fi, Digital Video Broadcast-Handheld (DVB-H), and MediaFLO. OFDM is a scheme that offers good resistance to multipath and allows WiMAX to operate in non-line-of-sight conditions.

At MAC level, WiMax offers two basic topologies: pointto-multi point (PMP) and mesh. The first mode resembles a cellular wireless network structure where each cell has its own Base Station that routes its traffic among its SSs. On the other hand, in the mesh mode, traffic may be routed through other SSs. Thus, a PMP is a centralized topology where the BS is the system center while in a mesh topology, it is not. Since the PMP topology is the option for the last mile wireless access, we consider it as the MAC topology in our proposed CAC method.

### System and Traffic Assumptions

The PMP topology illustrated prevails in the current analysis. In this array, the WiMax wireless link (represented by $B$ radio channels) is shared by SS (WiMax) and WiFi networks traffic flows. Each traffic flow fall into one the following service classes: WiMax real time connections, WiFi real time connections, and data connections.

Due to their delay-intolerant nature, WiMax and WiFi incoming request connections require constant bandwidths to fulfil their QoS profiles. Conversely, because of the elasticity, a data service tolerates variations in the service rate thanks to the TCP flow control mechanism. Additionally, as best-effort application it equally shares the residual bandwidth not used by WiMax and WiFi real time connections. In this respect, each data call service rate can change over time, depending on the number of ongoing WiMax and WiFi real time connections and data connections. Like, we do not differentiate WiFi data traffic from WiMax data traffic. The optimal CAC resides on the WiMax BS.

As usually assumed in performance evaluation of wireless networks, WiMax real time connections, WiFi real time connections, and data connections follow Poisson processes mutually independents with parameters $\lambda_x$, $\lambda_f$, and $\lambda_d$, respectively. Additionally, they require negative exponential service times with mean rates $1/\mu_x$, $1/\mu_f$, and $1/\mu_d$, respectively. We define $\rho_x = \lambda_x/\mu_x$, $\rho_f = \lambda_f/\mu_f$, and $\rho d = \lambda_d/\mu_d$ as the WiMax real time connection, WiFi real time connection, and data connection intensities, respectively.

## Proposed CAC Method

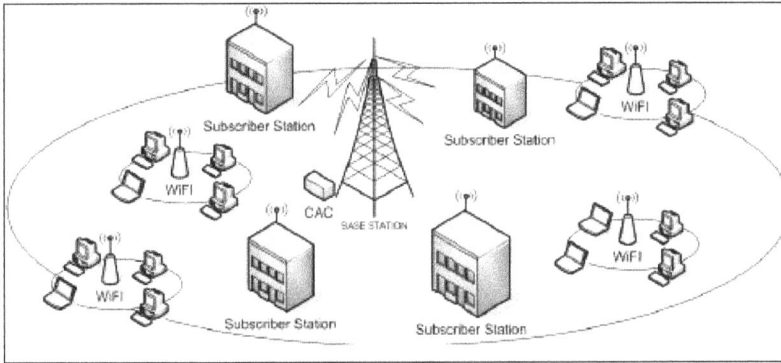

Scenario under analysis: WiMax and WiFi wireless networks integration.

Figure outlines the flowchart of the proposed CAC method. As can be seen, when an event is detected by the optimal controller, it first identifies the type of traffic class. It further determines the type of event, i.e. if a call departure or an incoming service request. This assessment is necessary because the optimal CAC determines the type of action to be taken based on this knowledge. Thus, for a real-time call arrival, it verifies, based on the proposed cost function (to be defined later), whether or not the incoming service request will be accepted and apply the data call preemption. In this case, the transmission parameters (ongoing bandwidth) of data calls are updated considering the admission of a new real-time call. On the other hand, for data service, the optimal controller always accepts the incoming service request as long as there is room to accommodate it according to its bandwidth requirement. Again, the data service transmission parameters are re-computed to conform the call to the residual radio re-sources. After the end of their services, real-time calls and data-services leave the system releasing their radio resources.

The proposed CAC method is quite general and cover the two major scenarios: i) when the WiMax Service Providers implement their own WiFi networks, which is a profitable business strategy; ii) when the WiFi networks are owned by third-party operators. In this case, a Service Level Agreement (SLA) must be established between the involved parties.

The proposal model does not cover the situation in which users migrate between WiFi and WiMax networks. In fact, we assume that they are stationary. This consideration represents the primary WiMax deployment goal in many regions around the world, mainly, in early stages in which the goals are the last mile wireless access with QoS support. A practical example of this is the Amazonian region. The lack of telecommunications infrastructure is latent in several cities. Additionally, the cities are, in many cases, surrounded by forest and lakes. These factors contribute to their almost total isolation and with them, their hospitals, schools, government offices, and so on. Currently, in many cities in such region, a WiMax network is being deployed aiming at, in this crucial moment, solely providing the last mile access. In a second moment, it is intended to interconnect all the cities with a fiber optical backbone forming digital cities. With this infrastructure implemented, several services are envisioned to be deployed.

## SMDP Model Formulation

### State Space

The set $\Phi$ of all feasible states is defined as:

$$\Phi = \left\{ \left( m_x, m_f, d, e \right) : 0 \leq m_x \leq \left\lfloor \frac{B}{Bx} \right\rfloor, \right.$$

$$\left. 0 \leq m_f \leq \left\lfloor \frac{B}{Bx} \right\rfloor, 0 \leq d \leq \left\lfloor \frac{B}{B_{\min}} \right\rfloor, e = \begin{bmatrix} 0 & 1 & 2 \end{bmatrix} \right\},$$

where $m_x$, $m_f$, and d denote the number of in-progress WiMax real time connections, WiFi real time connections, and data connections. Ongoing WiMax and WiFi real time connections require $B_x$ and $B_f$ radio channels to fulfill their QoS profile.

In the proposed CAC method, the preemptive priority provides resource assurance for (WiMax andWiFi) real time services over data calls. In order to prevent bandwidth monopolization of realtime services, we provide means of mitigating the preemption impact on the network performance. The first is degradation and compensation mechanism, which captures the elastic characteristic of data traffic. This way, a data connection can quickly finish its service by using as much radio resources as it can. Some works in literature have implemented this feature; however, it is unfeasible in practice to assume that a connection can use all the available bandwidth. It is our belief that it is more suitable to assume that the bandwidth allocated to a data connection varies within the following limits: minimal bandwidth ($B_{min}$) and maximum bandwidth ($B_{max}$).

In a nutshell, the degradation and compensation mechanism assumes that a data connection is accepted and served with the maximum bandwidth $B_{max}$ if possible. However, due to the resource dynamic occupancy, the CAC method will reassess the system load after any system state changes (motivated by call arrivals or departures) and adjust the actual bandwidth values between the minimal bandwidth and the maximum bandwidth accordingly. To model this traffic elasticity, the concept of ideal departure rate applies in which the real instantaneous departure rate of data connections is proportional to the actual bandwidth of each connection. So, with $m_x$ and $m_f$ real time connections into the system, each data connection will receive the bandwidth of:

$$b_w(i) = \min\left( B_{max} \, \max\left( 1, \frac{B - m_x B_x - m_f B_f}{d} \right) \right),$$

and will be served with service rate of:

$$\mu_{d'} = \frac{b_w(i)}{B_{max}} \mu_d.$$

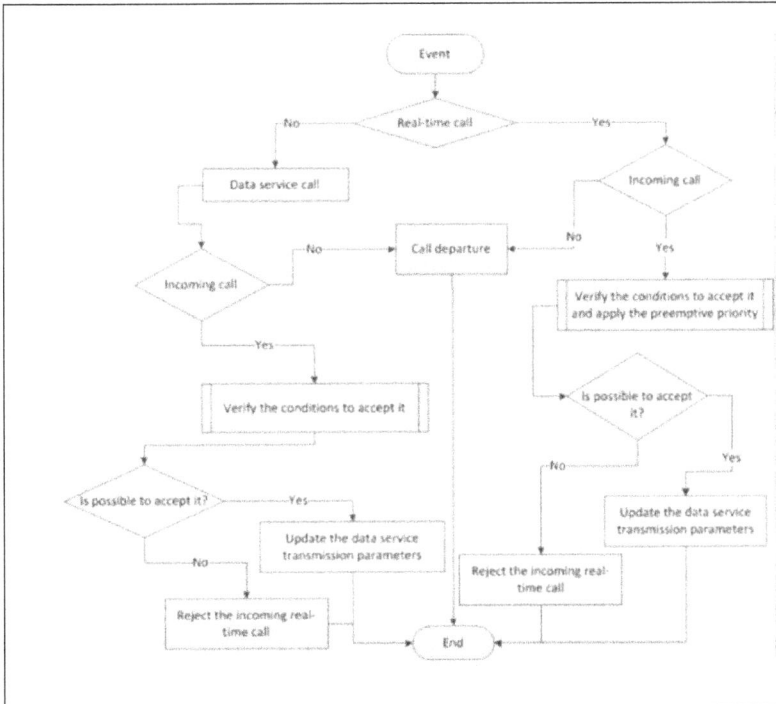

Flowchart of the proposed optimal admission control.

It is noteworthy to realize that inside the concept of ideal departure rate when a data connection receives the maximum bandwidth, $B_{max}$, its mean service rate will also be maximized and equal to $\mu_d = \mu_d$,

In equation, the random variable e is the last event occurred. This information is introduced in the state space in order to define the set of possible actions in each state. Accordingly the system dynamics, the values of e ∈ e may be: 0, 1, 2; where the former means either an arrival (departure) of data connection or a departure of a real time connections (WiMax or WiFi). The second means an arrival of WiMax real time connection and the latter means an arrival of WiFi real time connection.

## Decision Epochs and Actions

We assume that each state means the system's configuration just after an event occurrence and just before a decision making. The "real" decision epochs are the arrivals of real time connections, i.e., WiMax and WiFi real time connections; while the service completion epochs and arrival of data connection are defined as "fictitious" decision epochs, $e = 0$. In each state $i \in \ddot{O}$, the admission controller can choose one out of the possible actions:

$$A(i) = \begin{cases} a = 1, & e = 1, 2 \text{ and } \sigma \leq B \\ a = 0, & e = 0, 1, 2: \end{cases}$$

where $\sigma = B_j + \Sigma(m_x + B_x + m_f B_f)$ is the amount of bandwidth used by real time call plus the bandwidth required by the new connection request of type $j$, which is determined by the value of e = 1, 2.

In the set of actions $a \in A(i)$, $i \in$ , the action $a$=0 denotes the rejection, $a = 1$ denotes acceptance. After admittance, a WiMax real time connection or a WiFi real time connection may preempt the bandwidth being used by data calls or even some data calls. Since there is a minimum bandwidth requirement for a data call, it is needed to determine if the remainder bandwidth is enough to accommodate all the existing data calls. Therefore, after an admission, the remainder bandwidth can support:

$$\theta = \left[ \frac{B - m_x B_x - m_f B_f}{B_{min}} \right],$$

data calls with bandwidth Bmim. Thus, if d<0 then the system can support all the existing data calls with bandwidth more than Bmim; otherwise, $\zeta = d - \theta$ data calls will be preempted and the system will reduce the bandwidth of the remainder ones θ to Bmim. Mathematically, the number of data calls into the system after the admission will be given by min($d$, θ).

## Expected Time until the Next Decision Epoch

If the system is in the state $i \in Ö$, and the action $a \in A(i)$ is chosen, then the expected time until the next decision epoch, $\tau_i(a)$, is given by:

$$t_i(a) \frac{1}{\lambda_x + \lambda_f + \lambda_d + \Sigma m_x \mu_x + \Sigma m_f \mu_f + \Sigma d \mu_{d'}}$$

## Transition Probabilities

The state dynamic is completely specified by stating the transition probabilities among the system states. Thus, let $p_{ij}(a)$ be the probability that in the next decision epoch the state will be $j \in Ö$ if the present state is $i \in Ö$ and the action $a \in A(i)$ is chosen. For all $i$ and $j \in Ö$, we have the following cases:

$$Pij(a) = \begin{cases} \lambda_x \tau_i(a), & i = (m_x, m_f d, 1), \ a = 1, \\ \quad j = (m_x + 1, m_f, \min(d, \theta), e) \\ \lambda_x \tau_i(a), & i = (m_x, m_f d, 1), \ a = 0, \\ \quad j = i \\ \lambda_f \tau_i(a), & i = (m_x, m_f d, 2), \ a = 1, \\ \quad j = (m_x +, m_f + 1, \min(d, \theta), e), \\ \lambda_f \tau_i(a), & i = (m_x, m_f d, 2), \ a = 0, \\ \quad j = i \\ \lambda_d \tau_i(a), & i = (m_x, m_f d, 0), \ a = 0, \\ \quad j = (m_x +, m_f, d + 1, e) \\ \quad d < \left[ \dfrac{B - m_x B_x - m_f B_f}{B_{\min}} \right] \\ m_x \mu_x \tau_i(a), & i = (m_x, m_f d, 0), \ a = 0, \\ \quad j = (m_x - 1, m_f, d, e) \\ m_f \mu_f \tau_i(a), & i = (m_x, m_f d, 0), \ a = 0, \\ \quad j = (m_x, m_f - 1, d, e) \\ d \mu_{d'} \tau_i(a), & i = (m_x, m_f d, 0), \ a = 0, \\ \quad j = (m_x, m_f, d - 1, e) \\ 0, & \text{Otherwise.} \end{cases}$$

Due to the complexity of the proposed SMDP model, it is quite impractical to graphically represent a complete state transition diagram even for a small-scale system. Therefore, an example displaying the transitions for the states (1,3,8,0) and (1,3,8,1) $\in$ Ö is outlined. As shown, the system moves from the state (1,3,8,0) to state (1,3,8,1) or (1,3,8,2) upon an arrival of a WiMax real-time connection request or a WiFi real-time connection request with probabilities $\ddot{e}_x\,\hat{o}_i(a)$ or $\ddot{e}_f\hat{o}_i$, respectively. With probability $\lambda_d\tau_i(a)$,, the system moves from state (1,3,8,0) to (1,3,9,0). After its completion, a WiMax (resp. WiFi) real-time connection departures and triggers a transition from the state (1,3,8,0) to (0,3,8,0) (resp. (1,2,8,0)) with probability $m_x\mu_x\tau_i(a)$ (resp. $3m_f\mu_f\tau_i(\alpha)$). Similarly, a departure of a data connection with probability $8\mu_d\tau_i(a)$ motivates a state transition to (1,3,7,0). Since the event $e = 1$ is a decision making epoch, the optimal CAC has to select either the action $a = 0$ what implies in a connection rejection or the action $a = 1$ that triggers the transition to the state (2,3,8,1). By using the same approach, the remaining state transitions can be similarly obtained.

## Cost Function

In the proposed CAC method, the main goal is to find out a rule for maximizing the system capacity (minimizing real-time blocking probability) while controlling the application of the preemptive priority on data calls such that the long run average cost per time unit is minimal. To this end, the following cost function is proposed:

$$C_i(a) = C_x(i,a) + C_f(i,a) + C_p(i,a),$$

where $C_x(i, a)$, $C_f(i, a)$,, and $C_p(i,a)$ are the WiMax, WiFi real time connections blocking costs and preemption cost, respectively. These functions are computed as:

$$C_x(i,a) = \begin{cases} c_x & e = 1;\ a = 0 = 0 \in A(i) \\ 0 & \text{Otherwise} \end{cases}$$

$$C_f(i,a) = \begin{cases} c_f & e = 2;\ a = 0 = 0 \in A(i) \\ 0 & \text{Otherwise} \end{cases}$$

$$C_p(i,a) = \begin{cases} \zeta Cp_x, & e = 1;\ a = 1 \in A(i);\ \zeta > 0 \\ \zeta Cp_x, & e = 2;\ a = 1 \in A(i);\ \zeta > 0 \\ 0 & \text{otherwise} \end{cases}$$

where the quantities $c_x$ and $c_f$ are, respectively, the immediate cost incurred whenever an incoming WiMax or WiFi real time connection ($e = 1, 2$) is blocked.

In this topic, the preemption cost is employed as a manner to control the importance of real time services and data service accordingly the business model of the place where the WiMax/WiFinetworks will be deployed. Thus, the higher the $C_p(i, a)$, the higher the importance of data connection into the system and vice-versa. For the sake of simplicity, an intuitive definition of $C_p(i, a)$ is used. In our analysis, it assumed that it is proportional to the number of preempted data calls ($\zeta$), where $c_{px}$ and $c_{pf}$ are the immediate costs incurred whenever an incoming WiMax or WiFi real time connections is accepted and a data connection is preempted, respectively.However, it is noteworthy that this cost might depend on complex factors such as the signaling overhead and the amount of resource re-allocated needed to run and manager the preemption operation not only in the link under analysis, but also in others where the signaling overhead traffic is transmitted, network architecture (including the Operating System), and so forth.

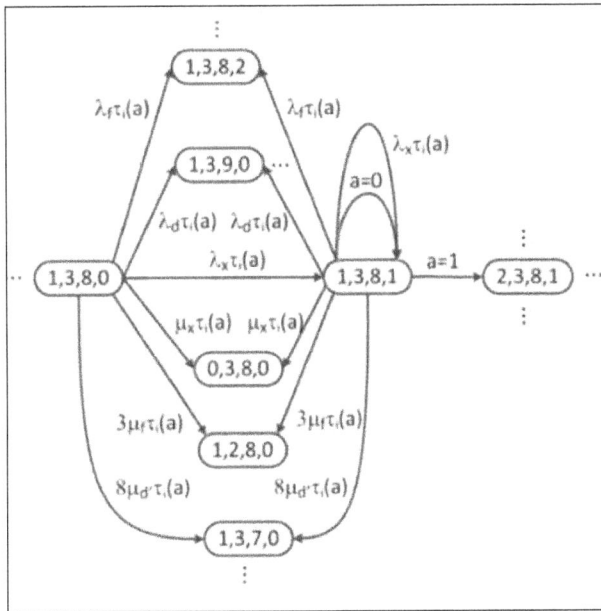

Transition Probabilities for the states (1,3,8,0) and (1,3,81).

## Data-Transformation Method and Value Iteration Algorithm

After the specification of the SMPD model, we have to convert it into a discrete time MDP model such that for each stationary policy the average cost per time unit in the discrete-time Markov model is the same as in the semi-Markov model. This approach is referred to as data-transformation method. Next, the value iteration algorithm may be applied in the transformed model to get the optimal policy. It is worth mentioning that a stationary policy $R$, defined by the decision rule f : $\Phi \rightarrow A$, prescribes the action $f(i) \in A(i)$ each time the system is observed in the state i $\in \Phi$ . The data-transformation method is described as follows: Let i, j $\in \Phi$ and $a \in A(i)$, choose a number $\tau$ such that:

$$0 < \tau < \min_{i,a} \tau_i(a),$$

and redefine the process according to the following procedure:

$$\overline{\Phi} = \Phi$$

$$\overline{A}(i) = A(i), \quad i \in \overline{\Phi}$$

$$\overline{C}_i(a) = \frac{C_i(a)}{\tau_i(a)}, \quad i \in \overline{\Phi}, \ a \in \overline{A}(i),$$

$$\overline{P}_{ij}(a) = \begin{cases} \dfrac{\tau}{\tau_i(a)} P_{ij}(a) & i \neq j, i \in \overline{\Phi}, \ a \in \overline{A}(i) \\[2ex] \dfrac{\tau}{\tau_i(a)} P_{ij}(a) + \left[ -\dfrac{\tau}{\tau_i(a)} \right], & i = j, i \in \overline{\Phi}^{\wedge}, a \in \overline{A}(i), \end{cases}$$

in which the notation $\overline{i}$ means the converted component. After turning the continuous time SMDP model into a discrete time MDP model, the value iteration algorithm, outlined in the Algorithm, is applied to obtain the optimal policy. When the value iteration algorithm stops, after finitely many iterations, the policy $R(n)$, whose average cost function is given by $gi(R(n))$, satisfies:

$$0 \leq \frac{g_i(R(n)) - g^*}{g^*} \leq \varepsilon,$$

where $g^*$ denotes the minimal average cost per time unit and $\varepsilon$ is the tolerance used to stop the algorithm.

## Algorithm Value Interation Algorithm

Require: $\Phi$, $p_{ij}(a)$, $A(i)$, $C_i(a)$, $\tau_i(a)$, $\tau, \varepsilon$

For each $i \in \Phi$, set $V_0(i)$ as $O_e \displaystyle\sum_{\substack{i \in \Phi \\ e=1,2 \\ a=1 \in A(i)}} \left( \lambda_x + \lambda_f + \lambda_d + m_x \mu_x + m_f \mu_f + d \mu_{d'} \right) \pi_i$ and set $n \leftarrow 1$.

$$0 \leq V_0(i) \leq \min_a \frac{Ci(a)}{\tau_i(a)}$$

For each $i \in \Phi$, compute $V_n(i)$ by:

$$V_n(i) = \min_{a \in A(i)} \left[ \frac{C_i(a)}{\tau_i(a)} + \frac{\tau}{\tau_i(a)} \sum_{j \in \Phi} P_{ij}(a) V_{n-1}(j) + \left( 1 - \frac{\tau}{\tau_i(a)} \right) V_{n-1}(i) \right]$$

Let $R(n)$ be a stationary policy whose actions minimize the right-hand side of equation above.

Compute the bounds:

$$m_n = \min\left\{V_n(i) - V_n(i) - V_{n-1}(i)\right\} and\ M_n = \max\left\{V_n(i) - V_{n-1}(i)\right\}$$

The algorithm stops with policy R(n) if $0 \le \dfrac{M_n - m_n}{m_n} \le \varepsilon$.

$n \leftarrow n + 1$ and go to first step.

## Performance Metrics

In this topic, we derive the QoS performance metrics used to assess the system performance. The mean carried real-time connection (WiMax or WiFi) traffic is computed as:

$$O_e \sum_{\substack{i \in \Phi \\ e=1,2 \\ a=1 \in A(i)}} \left(\lambda_x + \lambda_f + \lambda_d + m_x \mu_x + m_f \mu_f + d\mu_{d'}\right)\pi_i$$

where $\pi$ is the continuous time Markov chain steady state probability distribution vector under the optimal policy. Giving $O_e$, the real-time connection blocking probabilities are expressed as:

$$P_{brt} = 1 - \frac{O_e}{\lambda_e},$$

where $P_{brt}$ depends on the e value. Thus, if $e=1$ then $P_{brt}$ will be the WiMax real time connection blocking probability and $\lambda_e = \lambda_x$. The data connection blocking probability, Eq. is given by the probability that an incoming data connection faces less than the minimum bandwidth available.

$$P_{dc} = \sum_{\substack{i \in \Phi : d \ge \left\lceil \frac{B - m_x B_x - m_f B_f}{B \min} \right\rceil}} \pi_i.$$

The bandwidth utilization is defined as the ratio between the mean number of occupied channels and the total number of channels, i.e.

$$U = \frac{1}{B} \sum_{\substack{i \in \Phi \\ m_x > 0; \\ m_f > 0; \\ d > 0}} \left(m_x B_x + m_f B_f + dbw(i)\right)\pi_i.$$

The mean number of preempted data connections is given by:

$$N_{pd} = \sum_{\substack{i \in \Phi; \\ e=1,2; \\ a=1 \in A(i) \\ \zeta > 0}} \zeta \pi_i$$

# References

- Using-WiMAX-Technology-for-E-Learning-Solutions- 46486103: researchgate.net, Retrieved 15 May, 2019

- Mobile-backhaul-an-overview, futurenetworks: gsma.com, Retrieved 19 July, 2019

- Optimal-WiMax-backhauling-solutions-for-WiFi-traffic-283765010: researchgate.net, Retrieved 26 August, 2019

- Wireless telemedicine systems: an overview, IEEE Antennas Propag. Mag., vol. 44, pp. 143–153, Apr.2002

# PERMISSIONS

# INDEX

www.ingramcontent.com/pod-product-compliance
Lightning Source LLC
Chambersburg PA
CBHW061948190326
41458CB00009B/2821